PRAISE FOR
MOVING TO HIGHER GROUND

"John Englander has again written a timely, incisive, and important book. He explains, in clear language, the crises we are facing from sea level rise and offers positive ways to accommodate these challenges . . . a must read for all people concerned about our future."

—Christine Todd Whitman, former New Jersey governor and EPA administrator

"John Englander's new book takes a vital next step in discussing sea level rise. In simple yet memorable terms, Englander explains the science behind the shocking impacts to come. And discusses how to prepare for these inevitable, life-changing ramifications."

—Jeff Berardelli, CBS-TV News meteorologist and climate specialist

"A meter or more of higher sea level presents a profound global challenge for many professions and engineering will be at the sharp-edge of finding practical solutions. There is an urgent need to formulate an adaptive-ready, future-proofed approach to new build and retrofit, and we applaud Englander on the timeliness of his new book . . ."

—Dr Colin Brown CEng FIMechE FIMMM, CEO Institution of Mechanical Engineers, London

"*High Tide On Main Street* was on the 'Top 10' on my Commandant's reading list for the Coast Guard . . . From critical infrastructure, collapsed housing markets, displaced populations, and environmental refugees, the impacts are profound. In this new book, John Englander convincingly reorients our attention from the tyranny of the present to a future that absolutely must be addressed by national leaders. A 'must read' for climate change proponents and nay-sayers alike!"

—Admiral Paul Zukunft, US Coast Guard (retired), former Commandant

"In Hong Kong and South China, there is increasing awareness that sea level rise is a challenge. *Moving to Higher Ground* is timely, as people in Asia want to both mitigate and adapt to climate change. John gives us a perspective that this is a global issue with no boundaries and shows us the path forward."
—Dr. Christine Loh, Hong Kong University of Science and Technology;
(former) Undersecretary for the Environment, Hong Kong

"Oceanographer John Englander has done it again with a timely, very well-written and outstanding book on the greatest threat facing humanity today: climate change and sea level rise. His book recharges my batteries and gives me hope to tackle these challenges together. When we protect the ocean, we protect ourselves."
—Jean-Michel Cousteau, President, Ocean Futures Society

"A brilliantly written book that captures the unprecedented challenges the world faces as sea level continues to rise at rates unknown to modern humankind . . . The chapters are scientifically sound, and written with a wide readership in mind. It is a pleasure to recommend this book without any reservations.
—Dr. Robert (Bob) Corell, former National Science Foundation Assistant Director; Chair, U.S. program for climate change and IPCC; Nobel (shared recipient) climate change

"As an evangelist for science-based spirituality and theology, I am guided by reality and evidence. With his first book, John Englander became the prophet about the rising sea and how it fits into the story of the evolving Earth. His latest book, *Moving to Higher Ground*, raises the bar, inspiring us to adapt, to focus on future generations, and to consider our legacy."
—Rev. Michael Dowd, author of *Thank God for Evolution* and host *of Post-Doom Conversations*

"John Englander skillfully conveys the consequences of rising sea levels. From a military perspective, this will literally change the strategic landscape and challenge global defense forces. The melting Arctic has passed the tipping point. John advises us to prepare and think about bold solutions to these new challenges. Englander is a brilliant communicator, and his book is an enjoyable and enlightening read."

—Major General Kim J. Jørgensen, commander of
Danish Joint Arctic Command, Greenland

". . . John's newest book responds directly to a plea I heard from a US city mayor about the risks posed by sea level rise: 'We know we have a problem. We know we have to do something. Show us examples that make a difference!' *Moving to Higher Ground* shows us how to cope with the effects of SLR using 'intelligent adaptation.' Architecture and urban design are critical elements of this future."

—Thomas Vonier, FAIA RIBA. President UIA—
The International Union of Architects and former
President AIA—American Institute of Architects

"*Moving to Higher Ground* makes a compelling case that sea level rise, unrelenting and inexorable, will dramatically impact the engineering, construction, finance, and legal industries in the coming decades. Sea level rise will transform communities, ushering in the need for new legal concepts to cope with this challenge.

—Thomas F. Morante, attorney, Carlton Fields,
insurance and finance practice

"John Englander is at the forefront of alerting us about the dangers and costs of global sea level rise. His book bridges the critical gap between the physical sciences, the public, and policy and governance. He captures the challenges that face humanity with sea levels rising several meters over the next one or two centuries. I warmly recommend this fantastic book."

—Professor Eric Rignot, glaciologist, UC Irvine; Senior Scientist,
NASA Jet Propulsion Laboratory; Member, National Academy
of Sciences; Nobel recipient (shared) climate change

"Englander provides us fair warning. Disastrous collapse of the Greenland and West Antarctic ice sheets can be avoided only if the public pressures our leaders to rapidly phase out fossil fuels."

—Dr. James E. Hansen, renowned climatologist; former
director of NASA's Goddard Institute of Space Studies

MOVING TO

RISING SEA LEVEL
AND THE
PATH FORWARD

HIGHER GROUND

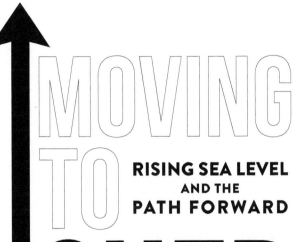

MOVING TO

RISING SEA LEVEL
AND THE
PATH FORWARD

HIGHER GROUND

JOHN ENGLANDER

FOREWORD BY **SENATOR ANGUS KING**
AFTERWORD BY **SIR DAVID KING**

Published by The Science Bookshelf, Boca Raton, Florida
www.thesciencebookshelf.com

Edited and designed by Girl Friday Productions
www.girlfridayproductions.com
Design: Paul Barrett
Project management: Sara Spees Addicott

Cartoons by Jim Bertram
© 2020 Jim Bertram

ISBN (hardcover): 978-1-7334999-0-3
ISBN (paperback): 978-1-7334999-1-0
ISBN (ebook): 978-1-7334999-2-7
ISBN (audiobook): 978-1-7334999-3-4

Library of Congress Control Number: 2020901623

First edition

This book is dedicated to everyone who rises to the bold challenge posed by the rising sea:

to the pioneers from whom I have learned parts of this incredible story and to all who hear the message, become teachers and "pay it forward."

CONTENTS

FOREWORD

SENATOR ANGUS KING

Time and tide wait for no man.
—English proverb (Chaucer)

When I taught management and leadership, one of the pitfalls we identified that undermines good decision-making is "the recency effect," the natural human tendency to assume that recent experience reliably predicts what will happen next. As much as we might expect the continuation of a hot streak, heads coming up five times in a row does not change the fact that on the next coin toss, the odds of either heads or tails are exactly fifty-fifty. Usually this concept is applied to truly recent experiences—as measured in minutes, hours, or maybe occasionally years. But what if "recent" happens to encompass all of human history?

Which brings us to sea level rise (and fall). It turns out that the unchangeable ocean we all know (especially those of us from places like Maine) isn't unchangeable at all and has varied constantly and significantly over time. It just happens that the last great variation took place right before humans started noticing (or at least keeping records of) such things. Fifteen thousand years ago, a moment in the age of the Earth, the sea that now laps at the shore of my hometown was 390 feet lower than it is today. You could have walked from Connecticut to the tip of Long Island; in fact, Long Island in those days wasn't an island at all; nor was Manhattan, for that matter.

In this extraordinary book, John Englander gives us valuable historical perspective, fascinating insights, and powerful data about what is happening now, and sound thinking about where we go from here (hint: it's most likely inland).

Fundamentally, there is a defined and constant amount of water on and around the Earth. Some of it is always in the atmosphere, but the vast bulk is in the oceans, rivers, and lakes of the world—and in the ice. Think of ice as stored water; more ice with less water, or less ice with more water. The period of low water between fifteen and thirty thousand years ago corresponded to the last great ice age, when glaciers covered much of North America, Europe, and the steppes of northern Asia. As the ice melted, we got Maine, and places like Scandinavia and Russia, and lots more water in the oceans.

There are two great frozen-water storage depots left as the climate has gradually warmed ("gradually" is an important word here, as we shall see later) and the glaciers melted—Greenland and Antarctica. To be precise, Greenland is storing more than 20 feet of sea level rise; Antarctica, almost 200.

I visited Greenland with John Englander several years ago and helicoptered over the vast ice sheet, which looked cold, forbidding, and permanent. Until we got to a series of enormous holes in the ice the size of a football stadium with rivers of blue meltwater cascading down into the depths, that is. Not only is that water eventually finding its way into the North Atlantic, it is also lubricating the interface of the ice and rock beneath, accelerating the gradual slide of the ice sheet seaward. We also visited the vast Jakobshavn Glacier on Greenland's west coast, which has retreated as far back into the ice sheet in the last ten years as it had in the prior hundred. (There's nothing like seeing for yourself to bring abstract scientific concepts like climate change into stark relief.) And this acceleration of the glacier's retreat is an important part of the message: We don't have thousands or even hundreds of years to confront this problem; we have relatively few years, and, if we're lucky, decades. During the melting of the great glaciers—about 15,000 years ago—there was a period when the sea level rose more than a foot each decade. As they say, it can happen here.

So all we need is a few more windmills and electric cars to reverse the Jakobshavn melting and close those holes in the ice, right? As John points out, the answer, unfortunately, is "no." Major steps to control fossil fuel use and build a carbon-free economy are necessary, but, alas, not sufficient. We have already passed the climate change tipping point; substantial sea level rise is now baked into our immediate future. But recognizing this reality doesn't diminish the urgency of our decarbonization efforts; we still have a chance to keep the costly and damaging from becoming the crushing and catastrophic.

In the pages that follow, John gives us a primer on how we got to this point and, more importantly, what we must do *now* to manage the flood that's coming at us.

Brunswick, Maine
April 2020

PREFACE

THE AIM AND AUDIENCE FOR THIS BOOK

This book is intended for those who want to understand and plan for the significantly higher sea level that is now inevitable. Audiences include:

- Homeowners, corporate executives, and investors with assets in the coastal zones including businesses, utility companies, industrial plants, and real estate developers.
- Professionals such as engineers, architects, transportation planners, insurers, lawyers, military, and members of nonprofit organizations who need to incorporate rising seas and shifting shorelines into their routine professional work and long-term planning.
- National and local policymakers who will have to change our most basic understandings of planning, national security, property rights, the public domain, liability, risk, and economics.

- Educators who will be revising a wide range of material to reflect the revolutionary new reality and future outlook.

The environmental consequences of sea level rise will be enormous and worthy of serious study. However, this book's primary focus is on the vast physical, practical, and economic impacts of unstoppable rising sea level, which will very likely be far worse than most can imagine. We can learn a lot from the period of Earth's history when sea level reached 25 feet higher (almost 8 meters) than today. That was part of the natural ice age cycles of the last 2.5 million years. The current meltdown will be shown to be something fundamentally different. This view on rising sea level has three purposes and goals:

1. To make clear the surprising height that sea level could reach in the coming decades, submerging most coastal areas rather permanently.
2. To present the immense challenge of adapting our civilization to an ever-rising ocean, in a way that appeals to our sense of being problem solvers and finding opportunities.
3. To encourage public urgency and support to slow the warming as much as possible, by transitioning off fossil fuels and using renewable energy sources and other technologies to reduce carbon dioxide levels.

Moving to Higher Ground updates and builds upon the science explained in my first book, *High Tide on Main Street: Rising Sea Level and the Coming Coastal Crisis.* Recognizing that many readers will not have read that previous book, I have included a few of its most effective visualizations and metaphors in this book. Everything in that book has stood the test of time and proven to be accurate. Since the first edition in 2012, many of the things described therein have become reality. In particular, many have noted a hypothetical scene I described, with extreme flooding devastating Atlantic City and New York City due to a hurricane on a certain path. With eerie timing on October 29, 2012, one week following the book's publication, Hurricane Sandy followed my scenario precisely. This book documents two other hypothetical situations I publicly described that also promptly became real. Facts can be stranger than fiction.

While dramatic sea level rise (SLR) seems scary, and may be impossible to imagine, the process is underway and is unstoppable in this century. Ice will continue to melt for a very long time on a warmer planet. The pace of SLR will accelerate, making flood events ever worse, moving most shorelines inland. Trillions of dollars of assets will literally go underwater.

Rising sea level will likely be the greatest agent of disruption and destruction this century, despite our best belated efforts to slow global warming. There will be huge problems, losses, and humanitarian challenges. But if we can see the situation with clear eyes and are prepared, there will also be tremendous opportunities in our transformation to a world with much higher sea level. It is urgent and imperative that we start planning and moving forward with awareness of the big picture. The geologic record makes crystal clear that sea level will be much higher than we ever imagined. That should reinforce the essential efforts to slow the warming, and to be more resilient to increasing short-duration flood events.

But a third issue is now independent of those two: We must begin *intelligent adaptation* to prepare for unstoppable rising sea level while there is still time to transition. While many view rising seas as just an environmental issue, I see it also as an economic and emotional one that can be used to engage and educate a wide cross section of society. For thousands of coastal communities, rising sea level will eventually become an existential challenge. It is my sincere hope that this book will help readers to be leaders in the amazing journey as we all begin the inevitable move to higher ground.

"I guess we'll have to move inland."

PART ONE

THE SCIENCE IS CLEAR; SEA LEVEL
WILL BE MUCH HIGHER

1

PAST THE TIPPING POINT—HIGHER SEA LEVEL, LESS LAND

Owing to past neglect, in the face of the plainest warnings, we have entered upon a period of danger. . . . The era of procrastination, of half measures, of soothing and baffling expedients, of delays, is coming to its close. In its place we are entering a period of consequences . . . We cannot avoid this period; we are in it now."
—Winston Churchill

In downtown Seattle in the late 1800s, flooding was getting really bad, particularly in the area now known as Pioneer Square, near the docks and Puget Sound. Back then, Seattle's problem was not rising global sea level. The land there was actually sinking, but that has the same effect of submerging property as would sea level rising. That low part of Seattle was built up with waste products from local lumber mills. Over decades, this fill land became soggy, rotted, and compressed, causing the ground to compact and the buildings to *subside*.

Land that moves downward is said to **subside** and increases relative sea level. Land that rises, or is **uplifted**, does the opposite. The vertical movement can happen violently during earthquakes, but usually occurs gradually due to compaction of sediments, pumping of water and oil from the ground, or the slow tectonic movement of Earth's crustal plates. Cities with extreme subsidence include Jakarta, Venice, New Orleans, and Norfolk (Virginia).

With the passing years, the flooding got worse, especially at the extreme lunar high tide, the phenomenon known as a spring tide, or the colloquial *king tide*. The monthly flooding got so bad that merchants and hotels were losing business and suffering property damage, even in good weather, following the varying tidal pattern. It was not just a problem of flooding streets; water was coming into some of the buildings despite the increased use of sandbags. Then came the great fire in 1889 that destroyed the area. With great foresight, the city took advantage of that disaster as an opportunity to fix the flooding problem. They committed to rebuild that area of the city 22 feet higher and started by raising the streets. For a few years, residents and businesses had to use wooden ladders to get up to street level and down to "the ground floor," which was now belowground. Over the course of a decade, the buildings abandoned their original ground floors and built new lobbies on what had previously been the first or second floors.

Those planners, property owners, and business leaders in Seattle realistically imagined a better future. They envisioned a solution that leapfrogged over their problem and looked ahead to the long term. The transition was expensive, unsightly, and disruptive. But their investment in a future-facing solution paid off in the long run. That decision to raise part of Seattle is a great example of adaptation, the central focus of this book. Today there are fascinating tours of "Underground Seattle" where you can visit blocks of those century-old structures, including hotel lobbies, now far below street level. A few have even been rehabilitated as underground coffee shops and bars.

As we look at the challenge of adaptation to rising sea levels and short-duration flooding in the coming chapters, there are a few other comparable examples to consider. Such solutions will not work everywhere. Many locations will lose value and eventually be abandoned, a process that has already started in some places. Other places can adapt and be protected in varying degrees, which will greatly increase their value. Our willingness to see the big picture will determine whether we invest wisely for the future as they did with Pioneer Square. To do that, we have to understand the cause and scope of the problem and where it is headed.

SEA LEVEL AND THE SHORELINE DEFINE OUR WORLD

Coastal flooding is getting noticeably worse all over the world. It's hard to miss the almost nonstop headlines. More extreme high tides, coastal storms, record rainfall, and rising sea level threaten where we live, our investments, national security, and even global security. The rising sea and migrating coastline get mingled with coastal erosion, though that's largely a separate problem. Ocean levels are rising higher than they have been in well over a hundred thousand years. Of all the factors that affect flooding, rising sea level is the phantom. Its effect is like a drip filling a bucket. Its signature is lost amid the extreme short-term flood events, but over time, its cumulative effect will raise the levels of all the oceans. From my global travels explaining rising sea level and the need to adapt, it is clear that there is widespread misunderstanding. In some cases it seems ideological, even political, but fundamentally it stems from misinformation about basic earth science and physics.

It's not uncommon for the topic of sea level rise (SLR) to provoke arguments about the problem and the solutions. Some dismiss sea level change and the broader climate crisis as *just a natural cycle*, ignoring the clear evidence that we have entered a new era of rapid rise that has broken out of the natural cycles and will not retreat for centuries. Others argue that reducing carbon dioxide emissions can stop sea level from rising further. Unfortunately, that's no longer realistic. Even some good scientists may poorly characterize the situation, by looking at something from a narrow perspective or with out-of-date information, adding to public confusion. We need to see the big picture and begin bold adaptation. There is no time to waste.

The shoreline is the most important line in the world, separating valuable real estate from that which is underwater. The location of the shoreline is determined by sea level. After about six thousand years of being stable, both are on the move. Shorelines are moving inland as sea level is moving higher. Not only will both continue inexorably for centuries, the rate is almost certain to speed up, greatly surprising nearly everyone. The change to our physical world will be hugely disruptive, completely unlike anything in our human experience. The disruption will extend deeply into the animal and plant kingdoms as sea level reaches heights unknown for over a hundred thousand years. Forecasts now predict sea level rising as much as 8 feet this century, with even further increases considered possible. Looking back in Earth's history, before human civilization, we can clearly see various times when sea level was hundreds of feet higher and lower than at present.

For millions of years, the amount of ice on our planet, sea level, and the shape of the land masses changed greatly. That seems hard to imagine, because all three have been quite stable for the roughly eight thousand years of human civilization. Earth has had many climate eras, some stable, some with patterns, and some with abrupt change. Past sea level changes left a clear record. Perhaps you have seen evidence like shark teeth and fossilized shells on land now well above sea level. After adjusting for land subsidence or uplift, we can get a very clear picture of how the ocean height has changed over time. The biggest factor for global sea level change is the amount of ice on land, in the form of the giant glaciers and ice sheets. Over centuries and millennia, as ice on land decreases, sea level height increases. The reverse is true as well.

Climate refers to long-term changes in weather characteristics such as temperature, precipitation, humidity, and winds, usually with thirty years as a minimum period to reduce the effect of any misleading spikes. Over centuries, the changing size of the polar ice and global sea level are proof of long-term temperature change.

Everyone knows that weather varies greatly year to year and can change violently, often with deadly effects. Yet, until recent decades, we thought that weather would average out over a period of centuries. Allowing for the year-to-year variations of major storms or heat waves, we thought it had long-term stability. With our advancing technology and science of the last half century, two facts about climate and weather have now come into mainstream acceptance in the scientific community.

First, even before the influence of humans, climate changed more abruptly than we thought. We now understand that Earth has shifted into warmer and colder eras—both the big ice age cycles, and the smaller eras of warming and cooling—at times faster, at times more slowly, and sometimes changing direction. Perhaps you have heard reference to the Little Ice Age or the Medieval Warm Period, examples of small variations in the highly complex Earth climate system. Major changes to the polar ice caps are among the best ways to define new climate eras. Dramatic changes to the frozen Arctic Ocean and the ice covering Antarctica and Greenland are now happening far ahead of expectations, breaking out of the natural cycles.

Second, the interaction of climate with the composition of the upper atmosphere—the realm of carbon dioxide and other so-called *greenhouse gases*—is more sensitive and dynamic than we previously understood. We are now seeing cascading effects as we get to *tipping points*. For example, the ice covering the Arctic Ocean is melting much faster than the models anticipated. It can be confusing even for scientists. The world and its climate are now changing before our eyes, in ways we never imagined, often rendering obsolete many of the textbooks from which we traditionally learned.

These two concepts are key to reconciling the changing climate of the past, prior to human influence, with the current era. The impact of nearly eight billion people is changing the composition of the atmosphere, the climate, the amount of ice, and sea level in a way that we find hard to recognize and even harder to accept. It can be very confusing, disorienting, and disturbing to consider the scale of this change, as the shoreline is a fundamental and seemingly permanent point of reference defining property and nations. The location of the coastline affects personal, commercial, and public property. Waterfronts and beaches draw tourism, recreation, and home buyers. Seaports, shipping, and fisheries are extremely valuable. Even the marshes and swamps are vital as the primary breeding ground for countless species. As sea level rises at faster rates, those ecosystems may have difficulty adapting.

Throughout the roughly eight thousand years of human civilization, the shoreline has remained a relatively fixed boundary. "Buy land—they're not making any more of it" has been an investment axiom passed on for generations, with the assumption that land is permanent, useful, and finite in quantity, making it the safest of all asset classes. However, as the rising sea encroaches on our shorelines at a quickening pace, coastal property is no longer *permanent*. Vulnerable property is subject to permanent devaluation.

THE "ICE AGES" UNLOCK THE SCIENCE

Our understanding of big changes in global sea level started to come into focus more than a century ago with the evidence for what is commonly called an "ice age." Though now perhaps best known through the five animated *Ice Age* children's movies, the underlying concept is rock-solid science. Mile-high ice sheets and glaciers extended far from the poles, into the midlatitude region. Those periods of expanded glaciers and ice sheets first occurred long before humans existed and are thus indisputably a natural phenomenon. As with any science, with each passing decade and better technology, we have refined our understanding, now recognizing many ice ages. The most recent one, properly known as the Pleistocene epoch, lasted approximately 2.5 million years. Within the Pleistocene, there were regular alternating periods of warm and cold, each lasting roughly one hundred thousand years. In common culture, we refer to each of the cold cycles as an *ice age*, which is how we will use the term in this book. With that narrow use of the term, the last ice age peaked about 20,000 years ago, entering the warming phase.

There is now good consensus about the cause for those regularly oscillating periods of hot and cold. A variation of less than one percent of the solar energy Earth receives is the trigger for these warming and cooling periods. The primary force for this phenomenon, known as the Milankovitch cycle, is our elliptical orbit around the Sun. When Earth is farther from the Sun, we receive less heat energy, similar to the way we receive less heat during the winter season each year. In fact, a good metaphor is that the hundred-thousand-year warming and cooling *ice age cycles* are like a large-scale version of summer and winter.

More information about **Ice Age Cycles and Causes** is available in Deeper Dive Note #1, at www.movingtohigherground. com. This is the first of ten online notes for those who want a little more science. You can read the Deeper Dive Notes as they are cited, or download the PDF for reference as a companion to this book.

In simple terms, when the Earth cools significantly over centuries, the ice sheets and glaciers at the poles, and on high-elevation mountains, become larger. That causes sea level to go down, since the oceans are the planet's primary water reservoir. As the oceans evaporate, the moisture in the air comes down as snow in the cold regions, slowly building up glaciers, causing ocean levels to go down over thousands of years. Conversely, as the planet warms, the ice sheets melt and shrink, with meltwater or glaciers entering the sea, causing sea level to rise. The point is that the amount of ice on land changes inversely to the volume of the oceans; one increases, the other decreases. Somewhat surprisingly, floating ice, such as icebergs, is different. As we will see shortly, floating ice does not affect sea level as it melts.

The most recent significant sea level rise began about eighteen thousand years ago following the natural warming pattern, coming out of the ice age cold period, causing the ice sheets to melt. For ten thousand years, sea level rose at an *average* of about 4 feet per century. As an annual average, that's only about a half inch, slightly more than a centimeter. The amount seems trivial and is impossible to observe directly given the variations of waves and tides. Yet like a drip filling a bucket, it's the steady accumulation year after year that has great effect. Ten thousand years of sea level rising at a half inch a year equals about

400 feet and is clearly confirmed in the geologic record. At that average rate, even in one hundred years, sea level would be 4 feet higher. Dealing with just 4 feet (1.2 meters) of sea level increase would be challenging for any coastal area. However, this century the rate of sea level rise will almost certainly be much faster than it was back then, because the current rate of planetary warming is much faster than in the natural cycles.

THE WORLD SHOULD BE COOLING; INSTEAD IT'S WARMING

The Pleistocene epoch ended 11,700 years ago at the start of the Holocene epoch, the period of relatively warm, stable climate lasting to the present. The Holocene could be viewed as the natural turning point within the ice age cycles. It was the relatively stable era in which all human civilization developed. Left to the natural patterns, over the next ten thousand years or so, the Holocene would almost certainly have turned into the next cooling era, heading toward the next ice age. Instead, we are now in this extraordinary warming period of the last century or so, tracking with the abnormal level of carbon dioxide, at least forty percent higher than in millions of years.

Thus, contrary to the widespread belief that we have accelerated climate change, we have actually broken out of the natural climate cycles characterized by the ice ages. The Holocene has given way to the Anthropocene, the new warming epoch characterized by human influence.

Seven or eight thousand years ago, at the end of that long period of dramatic rise, sea level reached roughly the present height, where it stabilized or *plateaued.* Many historians designate the start of human civilization as around 6,500 years ago, and perhaps even a few millennia earlier. It's intriguing that the rise of human civilization began at about the time that climate, sea level, and the coastline stabilized. Over millennia, without people understanding the larger pattern and mechanism of sea level moving up and down hundreds of feet, that stability became the prevailing belief that the height of the ocean was now settled and could be a reference point, like a surveyor's marker. As civilization developed, the height of something above or below sea level became a common measure and point of reference. It was presumed that the height of the sea and the location of the coastline were solid, like bedrock. That presumption was wrong. Rather than having permanent dimensions like granite, the height of the ocean and the placement of the coastline were changing quite like the dimensions of a glacier, which appears to be solid, but is slowly and constantly changing size and shape.

In the typical pattern of the ice ages, sea level repeatedly rises for roughly twenty thousand years and then falls for eighty thousand. The plateau phase is the turning point between the huge up and down swings of sea level. The recent six- to eight-thousand-year period, with rather "flat" sea level, was the end of the rising phase. According to the natural cycle, we should have now started the period of sea level falling. Instead, it's rising again. We have broken out of the long period of natural ice age cycles that has been in effect for millions of years. The breakout from the natural climate cycles of the ice ages correlates very well with the dramatically higher level of carbon dioxide in our atmosphere, which causes warming. The following graphic puts the rising sea level problem in clear perspective with global average temperature and carbon dioxide.

CO_2, GLOBAL TEMPERATURE, AND SEA LEVEL MOVE IN SYNC

This graphic depicts the four most recent cycles that we think of as "the ice ages" with a pattern repeating roughly every hundred thousand years. The graph starts on the left, four hundred thousand years ago, to the present on the right.

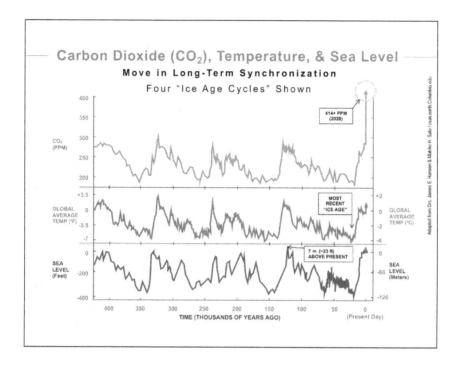

Figure 1. *The four most recent natural ice age cycles can be seen in the* **global average temperature** *record, the middle of the three lines in this graph. Approximately every 100,000 years, temperature, CO₂ (top line), and sea level (bottom line) cycle in close synchronization. Today, with the level of CO₂ above 400 parts per million (ppm), both temperature and sea level are rising, as expected, with a lag time. (John Englander graphic adapted from the work of Drs. James E. Hansen and Makiko Sato.)*

NOTE: To get all graphics in this book in full color, download the free PDF companion file from www.movingtohigherground.com. This graph is an important point of reference and will be cited in the following chapters, so you may wish to bookmark it.

Several key points about this graphic should be noted:

- Carbon dioxide, global average temperature, and sea level have changed greatly over millennia, with the three staying in very close long-term synchronization.
- On the middle line tracking global average temperature, it is easy to see the four most recent natural climate cycles, commonly described as *ice ages,* that have repeated approximately every 100,000 years for a few million years.

- The 100,000-year ice age pattern breaks into roughly an eighty/twenty split. The last cold point was approximately twenty thousand years ago. Our civilization of the last five thousand years or so has occurred at what would have almost certainly been the turning point, as we began eighty thousand years of cooling heading to the next ice age. That change of direction or turning point per the natural cycle gave us rather stable climate, ice sheets, and sea level.
- Now things are getting abnormally warm, breaking us out of the ice age pattern.
- The circled part of the top line, the level of carbon dioxide, has shot straight upward in the last two centuries, correlating with the burning of fossil fuels. (In Chapter Four, CO_2 and the correlation with temperature will be explained.)

The pattern of the three lines on the graph and some basic physics strongly suggest that temperature and sea level will continue to rise until the lines come back into balance. That will likely take centuries. The oceans are already warmer; the great ice sheets will melt for the foreseeable future. The atmosphere, the land masses, the oceans, and ice masses in the polar regions all warm at very different rates. There can be a lag time of centuries before the miles of ocean depth fully change temperature and the amount of ice reaches equilibrium, re-establishing stable sea level and shorelines. The lag time is confusing, but is key. An analogy with a lake may help. Imagine a large lake in a climate with warm summers and freezing winters. At the end of winter, with the air temperature above freezing, the ice will start to melt. Some larger blocks will remain, akin to miniature icebergs. With a few very warm days of spring, the surrounding land thaws, perhaps sprouting flowers. But the water remains near-freezing as long as any chunks of ice remain. The moment the ice has fully melted, the water temperature will start to increase rapidly. Even with a modest-sized lake, however, it could take many weeks for the water to warm enough to reach a new balance with the warming air temperature. The point is that land, water, and ice have very different heating properties. Returning to the global scale, the oceans are several miles deep, with enormous heat capacity. It can take centuries for them to reach a new state of balance as the planet warms. There is a huge lag time for the ice sheets and sea level to reach new equilibrium when long-term global average temperature changes.

Earth's climate is breaking out of the natural ice age pattern that prevailed for the past few million years at an explosive rate of change. Because there is no precedent for this in all of human civilization, it's hard for us to put in perspective and take seriously. Recent sea level rise and worsening coastal flood events only hint at what we will experience in the coming decades. With the rate of global warming now happening far faster than the natural cycle, the rate of sea level rise will very likely continue to increase for decades and centuries, partly depending on how quickly and effectively we can slow the warming. Regardless of efforts to slow the warming, however, the rate of sea level rise in the next fifty years will not be like the last fifty. Anyone planning for the future based on straight line or smooth curve extensions of the past will almost certainly be surprised.

WHAT DOES THIS MEAN FOR US, IN PRACTICAL TERMS?

Looking broadly across society and the impact on "us," this means that it's too late to focus our efforts exclusively on how to stop sea level rise by means of reducing greenhouse gas emissions and other good environmental efforts. We must begin to adapt now for the rising sea that is just getting started and will not stop anytime for the most basic reason. A warmer planet will have less ice. As the ice melts, the sea rises, pushing all coastlines inland.

It means that flooding in coastal areas will almost certainly continue to increase in frequency and water level, beyond anything in human history. "We" will be tempted to dismiss the extraordinary flooding as a freak event, or a trend that will soon reverse. Some will try to blame or associate it with things like sunspot cycles, or dismiss it as being like previous mini-ice ages and warm periods, all explanations that have been thoroughly discredited.

The sudden flood events, areas of coastal erosion, and extreme high tides are what will grab our attention. When there is flooding or even ponding water coming up from the ground, it will often be deeper, with a far wider area, and will appear more often. In addition to our understandable concern with those momentary high-water marks, we need to pay attention to the fundamental truth: the base water level is slowly rising, eventually causing permanent flooding or submergence. As others become alert to the sea change of increasing flood events, they too will consider protection or relocation before the flooding is permanent. The effect on property values will likely happen far sooner than expected, long before the property becomes submerged. Market valuations take into account perceived future risk. That applies to sophisticated risk markets as well as to our personal willingness to buy something. The difference from other market downturns is that when property values adjust downward for increased flooding, it is likely a permanent devaluation. As we will cover, there is simply no way for the aggregate ice sheets to increase and for sea level to start falling this century given that the ocean is measurably warmer.

While the rise in sea level and increased flooding will seem incremental, the acceleration, or rate of increase, can be deceiving. Due to some very simple physics, the rate of increase will accelerate and could become "exponential," which would surely surprise us. The implications increase by the decade. The effect on our children and future generations will be heightened. It means that low-lying homes and vacation spots on the ocean may not be passed on to future generations as might have been assumed. While this awareness may be depressing, it can be enlightening. Fortunately, with this challenge, we have time to adapt. It's important that we take advantage of that, overcoming the tendency to dither and delay.

"Global warming" is a term that I use interchangeably with **climate change** and is worth explaining. The planet is warming overall, which is ultimately the cause of the melting ice, the rising seas, and our inevitable move to higher ground. A few decades ago, it was realized that **warming** sounds positive to most people. Also, the term added to the confusion, as in certain times and places there were unusually cold temperatures and even more snow. Scientists and policymakers felt that **climate change** would be a better description. Yet that has proven to be too neutral or benign, not conveying urgency. The recent trend is to call it the climate crisis or climate emergency. In an ideal world, we would use a consistent term for clarity, but I don't see that happening soon, so I will use all of them at times, for both variety and nuance.

HOW MANY COMMUNITIES AND PEOPLE WILL BE AFFECTED?

A fallacy about rising sea level comes from pointing to places that are extremely vulnerable, such as Miami, New Orleans, Venice, or the Maldives. That communicates that the problem is largely limited to those locations and implies that other places are relatively unaffected. In reality, rising sea level is a real and present danger for nearly all of the world's approximately four thousand coastal cities, defined as having a population of 100,000 or more, as well as the far greater number of smaller coastal towns and rural areas spanning the globe. It is difficult to define a specific number due to the ambiguity of how to categorize small communities, but for the sake of simplicity, my first "SWAG" for this book—*scientific wild-ass guess,* a term often used in the military for an informed estimate—is that perhaps ten thousand coastal communities are at severe risk as sea level rises even 5 feet above present. I see little value in honing the number to be a few thousand higher or lower. The important thing is whether your community will be affected. Or whether the supply chains in your region, or the industries on which you depend, will be affected by the increasing short-term flood events and the creeping upward march of sea level.

The East Coast of the United States, from Maine to Florida and around the Gulf of Mexico, is particularly vulnerable to flooding due to the dominant gently sloping topography. For example, cities like Charleston, South Carolina, and Savannah, Georgia, are rarely thought of when SLR is considered, but are extremely exposed. The list of coastal places with high risk from even 5 feet of higher sea level is staggering. Internationally, there is vast exposure of population centers, from the giants of China and India, to low lying-Bangladesh and Vietnam, to diminutive Denmark. It's far more realistic to list the coastal communities that are not threatened by future inundation than to identify those that are.

Some coastal areas, like the Pacific Coast of the United States, have a generally sharp coastal rise, with most residents already elevated on cliffs and not nearly as exposed. Even there, however, rising waters will affect ports and harbors, with higher water levels moving far up tidal rivers. Global capitals, from Washington, DC, to London, are not coastal communities but are as affected by the daily rising tide as any coastal city and are very much exposed to rising sea level and worsening flood events. Sacramento, California; Hartford, Connecticut; and Hamburg, Germany might be better examples of usually overlooked cities far inland, on tidal rivers, that will need to deal with rising waters.

The question of how many people are vulnerable to flooding from rising sea level and short-duration flood events this century is even harder to define than the number of places. As we will cover in Part Two, the direct and indirect effects of accelerating SLR, combined with more flood events, make determination of a specific number quite unrealistic. Beyond the uncertainty of how much ice will melt in the coming decades, causing sea level to creep upward, and how bad the storms and rain will become, there is no objective way to define the number of buildings that will go underwater as the water rises inch by inch. How high does the water need to be for the value of a home to be destroyed? At a global level, just another inch higher sea level would put millions more people "over the line." Flooding combines rising seas, extreme tides, coastal storms, and storm surge, where higher ocean levels are caught in a confined basin, rising far higher than in the open ocean. The point is that even if the rising sea level reaches a particular height, that does not clearly define which properties are prone to flooding.

As important as the actual flooding is the trend. As the water moves ever higher at a quickening pace, the momentum and acceleration will affect attitudes about the need to relocate. At what point will a community's declining population trigger its economy to col-lapse, causing more people to move elsewhere? Location matters, too. Where are the people located? The sensitivity and ability to relocate will be different from Baltimore to Bangladesh and from Montauk to Mumbai. (I keep using different exemplar locations to remind us that SLR is not just "a Miami problem" as is commonly perceived.)

Depending on the criteria and the analysis, you will see diverse figures for the number of people subject to displacement from flooding this century. Calculations range from "as little as" 150 million up to a billion, the estimate recently given by the highly respected *Economist* magazine. It could be considered another SWAG. Regardless of the specific number, it is safe to say that hundreds of millions of people will be vulnerable to displacement by rising sea level combined with the short-duration flood forces of storms, rain, runoff, and extreme tides. The challenges will be enormous. The need to move to safe ground will be something quite new to us, and the discovery that the shore-line is no longer permanent will be deeply disturbing. As people look for where to resettle, there will be the looming question of how high and how far inland is safe enough. These troublesome, hard-to-answer questions will be a recurring storyline of this book as we consider the practical issues, including psychological, societal, environmental, eco-nomic, engineering, legal, and policy challenges.

As the rate of SLR accelerates, the waterfront will forever be changed. Yet a complete retreat from the shoreline is not an option. We will always need coastal access. In a changing world where coastal facilities and infrastructure are under threat from rising waters, the places that adapt smartly will become even more valuable.

Despite the protests, pontifications, and paralysis that the sub-ject invokes, we need to begin planning, designing, and building our new coastal future. We need to overcome our tendencies toward short-sightedness, procrastination, greed, and politicization. Even the many passionate environmentalists advocating for renewable or "green energy," pushing for recycling, getting plastics out of the ocean, con-cerned for wildlife, and advocating for humans to have a lighter foot-print on the planet, need to understand that preparing for substantially higher sea level needs to be done in parallel with all those good efforts. Understanding and planning for rising sea level should not be seen as competing with those issues, but rather as strengthening the case that we need to change the status quo.

Professionally, we are going to have to reinvent our approaches to engineering, architecture, land titles, easements, financial instruments, and even accounting to accommodate the rising sea and changing ocean coastline. *SLR is a threat to our economic well-being, the present world order, and international security. Adapting to much higher sea level is not optional; it will be the signature challenge of this century.* Yet, daunting as it is, if we can look over the horizon and plan for the new reality, future prospects are not entirely negative. While trillions of dollars of assets and economic value will literally go underwater this century, a comparable or even greater value can be created through the development of elevated or relocated housing, infrastructure, transportation grids, and utilities. New communities will have to be built. Some low-lying island nations will have to be accommodated elsewhere. Seeing that positive opportunity is an important goal of this book.

Our tendency seems to be to wait to act until the water is "waist-high" or until we have an accurate prediction for when it will reach a particular height. If we consider all the images of flood victims in recent decades in diverse communities, we know that the dumbest thing we can do is wait for the water to reach us—or for the timing to be specific and certain. The time to figure out "Plan B" is when there is still time to develop options. In other words, the time is now.

The global outbreak of the Covid-19 coronavirus in early 2020 demonstrates some striking similarities between pandemics and rising sea level. All major countries were warned for years about the risk of just such a pandemic. A few countries, like Singapore, South Korea, and Germany, appear to have been somewhat prepared. Most nations, including the United States, were not at all ready. There was a lack of equipment. There was no public health testing regime or even organizational structure ready to implement when weeks would make a big difference in the number of those who would get ill and die. As with the future threat of rising sea level, there had been warnings, but they were largely ignored. If there are common lessons, it is to listen to the scientists, to plan while there is time, and to develop necessary infrastructure before it is needed. By the time it's needed, it's too late to prevent catastrophe.

Whether your primary interest in adapting to rising sea level is in humanitarian issues; the challenge to plan, design, and build communities of tomorrow; preserving personal assets; or doing the best we can for future generations, we are now at the moment in human history when we must start *moving to higher ground*—metaphorically for most, and for many people on the coast, literally.

2

ANTARCTICA IS CRITICAL; MELTING ICEBERGS ARE IRRELEVANT

It ain't what you don't know that gets you into trouble.
It's what you know for sure that just ain't so.
—often attributed to Mark Twain

In case you're curious how I got into this, my fascination with changing sea level goes back to 1971 and a discovery I made 200 feet underwater. It was the summer before my senior year at Dickinson College, where I majored in geology and economics. A course in paleogeology introduced me to "the ice ages," more properly, *the Pleistocene.* That's when I first learned that *glacial periods* had occurred repeatedly over millions of years, with ice sheets a few miles thick covering the northern hemisphere, roughly down to the latitude of New York City. It was startling to know that geographic features like Long Island, Cape Cod, and the Great Lakes, which seemed so immutable and permanent, were created as recently as ten thousand years ago, remnants of the massive glaciers sculpting Earth's surface. Even more incredible to me was that, with each glacial cycle, sea level moved up and down hundreds of feet.

During semester breaks, I worked as a scuba instructor and dive guide at the Underwater Explorers Society (UNEXSO) on Grand Bahama Island. That summer I was teaching a deep diving course with special gear and procedures to safely go deeper than normal recreational scuba depths. We dove down along a vertical face of coral that dropped off from about 90 feet, almost straight down into the indigo darkness, stopping our descent at a depth of 200 feet. In that crystal-clear water, horizontal visibility was more than a hundred feet. Looking upward, sometimes we could see the boat far above, bobbing at anchor. What caught my eye on this dive was a flat, sandy area a few feet directly in front of me, terraced into the massive limestone face. It looked like a tiny beach amid the corals, sponges, and vertical limestone. I began to notice more areas just like it at the same depth. It seemed peculiar. Eventually it occurred to me that they might mark an ancient sea level.

Returning to college in the fall, I told my professor, who got excited and agreed it could be from when sea level was down that far during the most recent glacial period. Professor Hanson wanted some photos and samples from my next visit to the Bahamas, including sand, coral fragments, and shells, which I delivered. He said everything pointed to the same finding: Those sandy horizontal areas at the same depth showed where the sea level was about eleven thousand years ago. Apparently the sea, while rising 400 feet, paused at that halfway level for several centuries, allowing waves to create these miniature beaches. It was quite an aha moment. That discovery of ancient sea-level markers now 200 feet underwater made the ice age cycles very real to me. It would take a few decades for all the puzzle pieces to fall into place, but that was the start of my interest in rising sea level. Over the next few decades I did thousands of dives in diverse places. On several occasions, I went very deep in small submersibles. Amid all the fascinating things at depth, I would often note signs indicating the sea level and shoreline were at a lower level during the ice age periods. Basically, it was a matter of finding evidence for long-term wave action thrashing the shore. For me it was mostly just a point of interest and curiosity. Back then I never considered that sea level could change noticeably in my lifetime. It would be several decades before I awakened to the world-changing reality that sea level was again rising rapidly and that shorelines would be shifting inland, not in the usual geologic-time of millennia, but in real time; in our lifetime.

SEA LEVEL RISE IS HARD TO FATHOM

Confusion about SLR runs deep. Reasons range from our short time horizon in a world of *instant everything*, to generally poor geologic awareness, to special-interest-driven propaganda, intended to obscure the facts. Also, we simply do not want to believe the threatening message that sea level is on an unstoppable surge upward. The thought that the ocean could rise more than 10 feet (3 meters) is so deeply disturbing that we experience what psychologists call *cognitive dissonance*, essentially finding ways to ignore it. We often change subjects when confronted by a troubling topic. For example, I find that in conversations about the problem of SLR, people often change the discussion to the problem of plastics in the ocean—also a tough challenge, but completely unrelated to rising sea level. As you learn and consider the facts about SLR and its effect on the physical world, it's important to note how it affects us psychologically, including barriers to belief.

ICE ON LAND IS THE KEY TO RISING SEA LEVEL

Most people believe SLR is caused by melting icebergs and the melting polar ice cap, which is the vast ice around the North Pole. That's understandable, but both are floating ice. Intuitively, it would seem that as giant icebergs melt, the water level should rise, given that they stand high above the water, but a simple experiment proves this to be false. Fill a glass of water and add some ice cubes, but not so many that they touch the bottom of the glass. The ice cubes are like miniature icebergs. Mark the level of the liquid and let the ice melt. The water level will not rise. That's due to a very peculiar property of water. Though most substances become denser in solid form, water molecules become almost ten percent *less* dense just before they freeze into ice. A simplified explanation is that the rigid structure of ice crystals takes more space or volume than liquid water. Being less dense, ice cubes float almost ten percent above the water's surface. Conversely, when ice melts, the process reverses. The water molecules become denser and take up less space. The part of the ice cube sticking up above the water effectively disappears as it turns into the slightly more compact liquid water. Icebergs floating in the ocean are like the ice cubes, and in fact float a bit higher, since they are generally freshwater ice floating in the denser, saltwater sea.

There are three major forms of *floating* ocean ice: icebergs, ice shelves, and sea ice. Only about a tenth of the volume of each of these is visible above the water surface. Given the confusion about forms of ice and their effects on sea level and climate change, it's useful to distinguish them as follows and in the diagram below:

> *Icebergs* are very large floating masses of freshwater ice, broken off from a glacier or ice shelf. Lengths can be over 100 miles (160 kilometers).

> *Ice shelves*, also thick masses of floating ice, are distinguished by the fact that they build up where land meets water, particularly where glaciers meet the sea in a bay or cove. Thickness can range from 300 feet (about 100 meters) to 3,000 feet (about 1,000 meters). Because they float, they also do not directly affect sea level as they melt. Nonetheless, they have an indirect effect on SLR, often acting like corks in a bottle, holding back the glaciers on land, otherwise positioned to slide into the sea, which would raise global sea level.

> *Sea ice* consists of smaller areas of frozen saltwater, often less than 10 feet (3 meters) thick. Sea ice is very dynamic and changes with temperatures and ocean currents. It can be vast flat areas or piled up in huge blocks. Sea ice is estimated to cover twelve percent of the ocean, though that figure is decreasing quickly as the oceans warm.

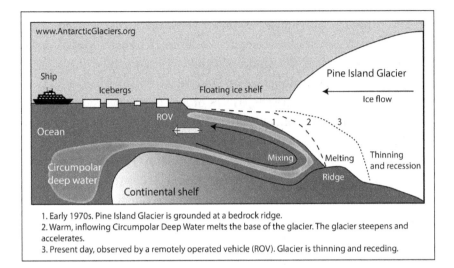

www.AntarcticGlaciers.org

Ship · Icebergs · Floating ice shelf · Pine Island Glacier · Ice flow · ROV · Ocean · Circumpolar deep water · Mixing · Melting · Thinning and recession · Ridge · Continental shelf · 1 · 2 · 3

1. Early 1970s. Pine Island Glacier is grounded at a bedrock ridge.
2. Warm, inflowing Circumpolar Deep Water melts the base of the glacier. The glacier steepens and accelerates.
3. Present day, observed by a remotely operated vehicle (ROV). Glacier is thinning and receding.

Figure 2. Cross-section illustration of Pine Island Glacier, one of the largest in Antarctica, to help distinguish the glacier, the ice shelf, and icebergs. Note how the ice is being eaten away on the underside; the distance is tens of miles. When this single glacier fully slides into the sea, global sea level will rise approximately 1.5 feet (0.5 meters). It is not possible to accurately predict when that will occur due to the complex and enormous forces involved and unknowns such as the actual rate of planetary warming. (Graphic courtesy Bethan Davies, www.antarcticglaciers.org.)

Like icebergs, sea ice and ice shelves are floating, or at least their mass is supported by seawater. Therefore, for the same reason that icebergs do not add to water level as they melt, floating sea ice and ice shelves do not either. They are each roughly ninety percent below the surface for the same reason icebergs are.

It's important to recognize that the North Pole is ocean, not land. Some maps show the Arctic Ocean as blue, but many show it as white, representing year-round ice. The Arctic Ocean has been frozen for most of the last three million years. Though that gives it the appearance of being a landmass, the entire area is actually floating sea ice. For most of human civilization, the ice has been relatively static in size and seen as a single mass, *the polar ice cap*. Now the floating ice is breaking up and disappearing very quickly as the Arctic warms even faster than the global average warming. Because the vast area around the North Pole is melting sea ice, it is totally different from Greenland, which is mostly ice on top of land, like Antarctica.

The primary contributor to higher sea level is the melting of *ice on land*, which essentially exists in two forms: *ice sheets and glaciers*. The two biggest ice sheets, Antarctica and Greenland, hold about ninety-eight percent of the global ice on land, representing over 200 feet of potential SLR. Glaciers are the "rivers of ice" that move slowly across land. A glacier is commonly visualized as distinct, perhaps in a ravine or valley between mountains in diverse places like the Alps or Alaska. In the two vast ice sheets covering Antarctica and Greenland, there are also thousands of glaciers embedded within ice sheets. Those glaciers are difficult for non-experts to identify but are the greatest source of potential SLR.

Such additions from ice on land are the big factor for SLR. There are three ways that these glaciers increase sea level: a) breaking off and falling into the sea, known as "calving" a new iceberg, b) slowly sliding out onto the water, moving from land to be a floating ice shelf, or c) as meltwater that finds its way to the sea.

Returning to the demonstration above about ice cubes melting in a glass as proxies for icebergs melting may help to make this clear. Adding water to the glass or adding a new ice cube will obviously raise the water level. That symbolizes the meltwater from ice on land, new icebergs breaking off from a glacier, or a glacier sliding into the sea. All those will cause global SLR at the moment they enter the sea. Once they are in the ocean, they will not have further impact on SLR, with one exception: as the water expands as a result of warming.

Thermal expansion of seawater has been the second-largest contributor to global sea level rise, behind melting glaciers and ice sheets. Water increases very slightly in volume as it gets warmer. Over the last century, as the oceans have warmed almost 2° Fahrenheit (1° Celsius), they have increased in height by about 4 inches (10 cm) just from the thermal expansion. This contribution was almost equal to that from melting glaciers, but the ratio is now changing. Looking ahead, thermal expansion will continue to add to sea level and even increase but will soon be greatly overshadowed by the increasing melt rate of glaciers and ice sheets.

Glaciers typically advance towards the sea on the order of a few feet per day, about a meter, though some are now moving thirty or forty times faster. Giant glaciers, which can be hundreds of miles long and up to two miles thick, are propelled by gravity, their massive weight, and melting. Glaciers normally grind their way slowly on top of bedrock. On their journey to the ocean, they bend and stretch. The warmer the temperature, the faster they make their way around hidden hills and down valleys towards the sea. Their path may be steered by land contours, but glaciers go where they want to go. No man-made structure can stop glaciers. To give a sense of scale, the enormous US Great Lakes were essentially "nick marks" just a few hundred feet deep, caused by mile-high glaciers about ten thousand years ago. In the new era of rapid warming, meltwater creates a new dynamic as it gets under the glaciers, finding its way down to the bedrock far below. The meltwater becomes substantial streams and rivers, with the glaciers effectively floating on top. Even very cold water is extremely effective at melting ice. Due to the warming and softening of the glaciers and the lubrication of the water underneath, some of the largest glaciers are now moving toward the sea at double, triple, and even quadruple their speed just a few decades earlier. Over 200,000 glaciers are now identified and tracked globally. With very few exceptions, all are now shrinking.

GREENLAND IS COLLAPSING IN REAL TIME

In recent decades, scientists studying the Greenland ice sheet have witnessed an alarming rate of increased melting and of icebergs calving off into the ocean. Measurements are done by satellite, airplane, and research stations out on the vast ice sheet. The Greenland ice sheet is collapsing in real time. Jakobshavn, the biggest glacier on the island, meets the sea near the town of Ilulissat. Jakobshavn spawns the largest number of icebergs in the northern hemisphere. A particularly stunning collapse of the glacier was caught on video in 2008 when a documentary crew was filming continuously around the Arctic with a few dozen cameras for the film *Chasing Ice*. In seventy-five minutes, they just happened to record a mass of ice comparable to the size of Manhattan calving off from the glacier.[1] In the decade since, other similar collapses have happened in the same area, but without a video crew pre-positioned to capture the event.

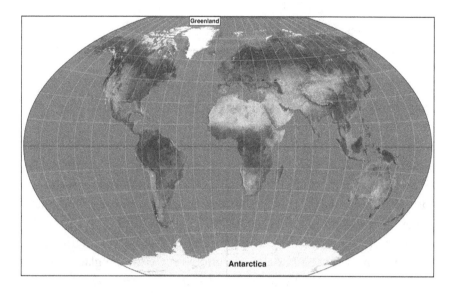

Figure 3. If all the ice on Greenland and Antarctica melts, global sea level would be more than 200 feet (60 meters) higher than present. Those two areas contain about ninety-eight percent of the total ice on land and are the dominant factors influencing future sea level rise. Both are melting faster than projections. Once we recognize the massive significance of Greenland and Antarctica to coastlines all over the world, these desolate and inhospitable regions take on a whole new significance. (Image: Winkel triple projection—Wikimedia Commons derived from NASA.)

ANTARCTICA: THE BIG FACTOR FOR SEA LEVEL RISE

Antarctica is many times larger and has seven times more ice than Greenland, and therefore, has that much more potential to contribute to SLR. Its larger size and different weather patterns make Antarctica much slower to melt than Greenland. In fact, until the last few years, Antarctica appeared not to be adding to SLR at all. For a while, East Antarctica—about two-thirds of the total continent—appeared to be getting larger in certain dimensions. Mostly it was the fringing sea ice, expanding the total area of Antarctic ice coverage but irrelevant to sea level, since it floats. Nonetheless, some skeptics took this phenomenon as evidence that climate change was not affecting Antarctica. In the last decade, the warning signs from Antarctica have become crystal clear. Not only are the vast ice shelves calving icebergs up to a hundred miles in length, there are ominous signs from all quadrants of the frozen continent. Dire warnings about the potential for Antarctic collapse resulting in catastrophic sea level rise go back to 1978, when the late glaciologist Dr. John Mercer wrote:

> *If present trends in fossil fuel consumption continue, and if the greenhouse warming effect of the resultant increasing atmospheric carbon dioxide is as great as the most advanced current models suggest, a critical level of warmth will have been passed in high southern latitudes 50 years from now, and deglaciation (reduction) of West Antarctica will be imminent or in progress. Deglaciation would probably be rapid once it had started, and when complete would have led to a rise in sea level of about 5 meters (16 feet) along most coasts.*[2]

Since we are now in the fifth decade after his fifty-year warning, we should be able to see if he was correct or not. Amazingly, just as I was researching and writing this book, a featured article in the weekly magazine of the American Geophysical Union (EOS, Feb. 17, 2020) caught my eye:

Glacial Earthquakes First Seen on Thwaites

Thwaites is the dominant glacier in West Antarctica, the focus of Mercer's concern, now often referred to as "The Doomsday Glacier" for its potential global effect on coastlines. The date of that article was February 17, 2020—forty-two years after Mercer's fifty-year fearful forecast. His prescient scenario is happening in the headlines. "Imminent or in progress" were his words, quoted above. In the realm of geologic and glacial timescales, Mercer's forecast has been validated. Because this one glacier is so critical and seen as an indicator for the other mega-glaciers, there are many studies in progress, particularly a five-year joint British-American project. One of its early findings in 2020 was the discovery of warmer seawater in a cavern nearly the size of Manhattan underneath Thwaites.[3]

To put Thwaites Glacier in scale, it is comparable to the size of Florida. Gravity will take it into the ocean sooner or later, at which point that single glacier will raise global sea level more than one and a half feet (about half a meter). Fortunately, it cannot happen suddenly like a tsunami, earthquake, or avalanche. Unfortunately, however, as with those abrupt events, there is no way to accurately predict when a major collapse will happen. And there is no reason to believe the collapse will be limited to a single glacier. As can be seen in the lower left of the map just below, Thwaites is grouped with others in West Antarctica, often referred to as the Pine Island Glaciers. Just as Mercer hypothesized, it is very likely that these particular glaciers hold the key to when the entire world will experience several meters (roughly 10 feet) of higher sea level.

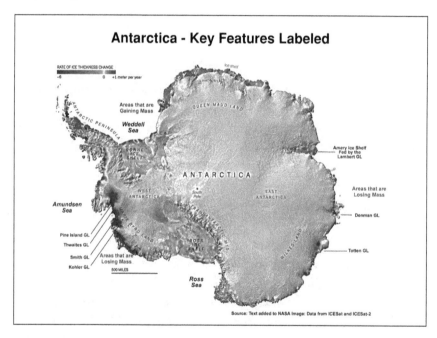

Figure 4. Different parts of Antarctica have vastly different potential to affect sea level rise. By themselves, the huge glaciers draining into the Amundsen Sea will add as much as 10 feet to global sea level as they eventually slide into the sea. It is not possible to accurately predict when that will occur. Recent measurements show the Thwaites and Pine Island Glaciers are showing significant signs of movement.

West and East Antarctica are both changing rapidly, as is Greenland. Until recently, East Antarctica seemed quite stable. By 2020 two of the largest glaciers there, Totten and Denman, were showing early signs of movement. For more specific and current information, see the online Deeper Dive Note #2 **Antarctica and Greenland Update**, at www .movingtohigherground.com.

MISLEADING JOURNALISM

Unfortunately, even some good science journalism can misinform. For example, a January 2018 article in the usually responsible *Washington Post* was titled:

> *Huge Snowfall Increases Over Antarctica Could Counter Sea Level Rise, Scientists Say*

It reported that increasing snowfall on East Antarctica could lock up more water on land, which, in principle, would lower global sea level. Careful reading of the article, however, revealed this effect might lower global sea level by 1.5 millimeters by the end of the century. That's only six hundredths of an inch, the thickness of a few sheets of paper, spread over eight decades. Perhaps scientifically interesting to someone, but completely irrelevant from a practical standpoint. Worse, that headline suggests that we might not need to worry about SLR, playing to our fantasy wishful thinking for a magical solution, adding to the delay and avoidance of real adaptation.

Another equally misleading story recently got into mainstream media. A graduate student at a prestigious university proposed building barriers to stop Thwaites and the other monster glaciers from their advance towards the sea.[4] Nonsense! These glaciers are not like the modest ones you may have seen in Alaska, the Alps, or Glacier National Park. Imagine a glacier, miles high, the size of Florida, that is moving downslope toward the ocean. It is delusional to think that humans will in any way deter its path. Encouraging students and others to come up with creative ideas is great, but taking such sophomoric ideas seriously without experienced engineering evaluation is irresponsible. Writers, editors, and readers need to be on guard for attention-getting headlines and articles that mislead, either due to an academic seeking attention, an editor looking for *sensationalism*, or those practicing propaganda to undermine concern and action about the warming planet.

Ice Core Sampling. A common question is: How do we know the history of climate change and the ice sheets before modern instruments and record-keeping? Ice core analysis began in the 1990s and is now a mainstream technique to unlock the secrets of annual climate variations by examination of drilled ice samples from Greenland and Antarctica. The annual layers of snow compacted into those ice sheets contain encapsulated air bubbles dating back as far as 800,000 years, the oldest ice sheets with a continuous record. In addition to providing annual samples of carbon dioxide levels, these samples document with remarkable precision how temperatures have changed over time. Amazingly, there is a temperature record, stored in the air bubbles by the ratio of oxygen isotopes. The technique can be thought of as similar to how we use tree rings to look at annual history for thousands of years.

ALASKA, ALPS, SCANDINAVIA, ETC.

Beyond Antarctica and Greenland, the remaining two percent of the world's land ice is spread across regions such as Alaska, the Alps, and the Himalayas, along with snow and ice in Canada, Russia, Scandinavia, the Western United States, and isolated glaciers from Africa, Peru, and New Zealand. If you have seen any of them, you know they can be enormous. Yet, even if all the snow and ice from those areas melted, global sea level would only rise a few feet (about a meter), less than two percent of the global total for potential SLR, which is estimated to be about 212 feet (65 meters). It seems hard to believe, but many estimates from different sources are in rather close agreement on that calculation. It's not that those glaciers and snow-shrouded mountains are small, but rather that the ice mass of Antarctica and Greenland is so gigantic. A quick look at where the potential two hundred feet of global sea level rise is being stored as ice tells the story:

Location of Potential Sea Level Rise	Feet	Meters	Percent
East Antarctic Ice Sheet	169	51.6	80
West Antarctic Ice Sheet	15	4.5	7
Antarctic Peninsula	2	0.5	1
Greenland Ice Sheet	24	7.3	11
All Other Glaciers	2	0.6	1
Total Potential Sea Level Rise	**212**	**64.5**	**100**

Adapted from Allison et al., 2009[5]

SEA LEVEL DOES NOT RISE SMOOTHLY

The following simple graphic of sea level rise since the last ice age peaked about twenty thousand years ago makes clear that sea level does not rise in a straight line or even a smooth curve. There are *bumps* as the polar ice sheets go through stages of collapse. As can be seen, ten thousand years ago, sea level rose suddenly tens of feet in a few centuries, indicated by the three arrows. The steepest segment, starting with the first arrow, had SLR of 65 feet (20 meters) over five centuries.[6] Though that only averages an inch and a half a year, over a century, that's approximately 13 feet of global SLR in a hundred years, a staggering rate of rise. The point is that collapse of the great ice sheets, the major cause of SLR, has accelerated suddenly in the past and will almost certainly do so again in the future.

Floods of "Biblical Proportions"
Many scholars believe that the epic floods described in such
diverse sources as the Book of Genesis, the Quran, the Epic
of Gilgamesh, the Hindu story of Manu, and the oral history
of the Haida, may have a basis in the sudden sea level rise of
more than thirteen feet (four meters) per century, ten thousand years ago.[7]

Next to the simple graphic showing how sea level rose since the
last ice age is a forty-seven-story building to help visualize the scale
of actual sea level. Imagine that twenty thousand years ago, when the
great ice sheets largely covered the northern hemisphere, sea level was
at the ground floor of this building. Using a virtual elevator, as the
great ice sheets melted over about fifteen thousand years, sea level rose
some 400 feet to the thirtieth floor, where it has remained for the last
six thousand years or so of human civilization. With the warming of
the last two centuries, sea level and our virtual elevator slowly moved
up about 10 inches (25 cm). Just in the last few decades, scientists on
our "elevator" have noticed a subtle but significant increase in the vertical speed upward. If we allow the planet to keep warming, melting all
the ice, sea level would rise about 212 more feet (65 meters), up to the
forty-seventh floor.

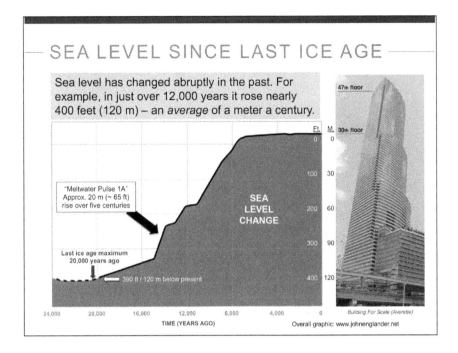

SEA LEVEL SINCE LAST ICE AGE

Sea level has changed abruptly in the past. For example, in just over 12,000 years it rose nearly 400 feet (120 m) – an *average* of a meter a century.

"Meltwater Pulse 1A" Approx. 20 m (~ 65 ft) rise over five centuries

Last ice age maximum 20,000 years ago

390 ft / 120 m below present

SEA LEVEL CHANGE

47th floor
30th floor

Ft. M.
0   0
100 30
200 60
300 90
400 120

Building For Scale (Averette)

24,000   20,000   16,000   12,000   8,000   4,000   0
TIME (YEARS AGO)

Overall graphic: www.johnenglander.net

Figure 5. Since the peak of the last ice age 20,000 years ago, sea level rose almost 400 feet (120 meters), at times abruptly, reaching the present level about five thousand years ago. That is equal to the thirtieth floor of a typical commercial building. If all the remaining ice melts, global sea level would rise another seventeen floors, 212 feet (65 meters).

The warming planet has already committed us to considerable sea level rise due to the ice sheets and glaciers melting, similar to what happened in the past, but much faster. To close this chapter with the metaphorical building, it is no longer possible for us to stay at the present sea level, the thirtieth floor on our virtual elevator. We are headed to the next "floor" some ten feet higher. Without extraordinary and urgent efforts to halt the warming, our virtual elevator and real-world sea level are headed *much* higher.

3

WHY SEA LEVEL RISE IS UNSTOPPABLE AND UNPREDICTABLE

All models are wrong, but some are useful.
—George Box

My core case for bold, visionary adaptation to accelerating SLR rests on three key concepts that are often misunderstood or ignored:

1. Sea level rise this century is unstoppable.
2. Science cannot precisely predict the rate of rise.
3. The rate of rise can accelerate quickly, possibly abruptly, greatly surprising us.

Rising sea level is happening so incrementally that it's easy to ignore. We are lulled into thinking we can prepare for the ocean rising when we know more precisely when it will happen. As with earthquakes and pandemics, the time to prepare is far in advance. Adding to the problem is that there are so many conflicting estimates and so much misinformation. Because we have never experienced much higher sustained sea level, we have great difficulty imagining it and taking it seriously. Before wrapping our minds around the challenges of coastal reengineering, we must understand what impact global sea level could have in the coming decades, including the extreme scenarios that are rapidly becoming more realistic possibilities.

WHY SEA LEVEL RISE IS UNSTOPPABLE

You now appreciate that global sea level moves up and down hundreds of feet vertically, mostly reflecting the size of the two great ice sheets on Antarctica and Greenland. Obviously, the size of the ice sheets changes with the planet's overall temperature. The oceans are already about 1° Celsius warmer over the last century, roughly 2° Fahrenheit (1.8°, to be exact). That may sound trivial, but it's not. For perspective, over the range of the ice age cycles, global average temperature moves up and down approximately 9° Fahrenheit (5° Celsius), as shown on the chart in Chapter One, Figure 1. If you consider the size of the oceans and their average depth of a few miles, it becomes clear that it takes truly unimaginable amounts of heat to warm the world's ocean an extra degree. Because it is so much denser, water absorbs and stores heat far better than air and maintains temperature stability, making it the preferred place to gauge planetary temperature changes. Ninety-three percent of the excess heat being trapped in the atmosphere by greenhouse gases is stored in the sea.

How much heat are we adding to the oceans and to the planet overall to raise the temperature level one degree? Since the measurement in watts, joules, or calories would be gigantic and a rather meaningless number to most people, scientists have calculated it in terms of a unit of energy that the world can visualize: the atomic bomb used by the US to end the Second World War, equal to about thirty million pounds of TNT explosive. Regardless of one's position about the devastating use of the weapons on Hiroshima and Nagasaki, the world saw the power of a single bomb to destroy an entire city. The infamous mushroom cloud has come to symbolize a new level of power. Here, we just want to use the heat energy from a single one of those bombs as a unit of measure. The increased level of greenhouse gases, primarily carbon dioxide, is now trapping extraordinary heat in the Earth's system, roughly the equivalent of 500,000 atomic bombs every day. That's about five atomic bombs being detonated per second, twenty-four hours a day, seven days a week.[8] That extra heat energy is warming the oceans and the atmosphere, melting the ice, and raising global sea level. It will continue to accelerate as long as we keep adding carbon dioxide and other greenhouse gases to the atmosphere.

Even if we could immediately stop all carbon dioxide emissions and the warming, the excess heat already stored in the sea will continue to melt the ice sheets for centuries, raising global sea level to heights unknown for the past hundred thousand years. On our primary graph in Chapter One, you can see the previous warm period of the ice ages (last *interglacial*) peaked 122,000 years ago. Global sea level reached about 25 feet (about 8 meters) above present. That warming was natural, unaffected by humans. Even without the present-day abnormal warming, that natural high-water mark should be sufficient cause to take seriously the need to begin our move to higher ground. Though it may take centuries to reach such heights, I have never heard anyone make a sober case that we can avoid sea level becoming significantly higher.

For those who want additional explanation about **why SLR is unstoppable**, the online Deeper Dive Note #3 (www.movingtohigherground.com) explains:
- Melting ice, warming oceans, and rising seas have strong force to accelerate, the "snowball effect"
- Why cooling the oceans is not realistic
- How Earth's balanced incoming and outgoing energy of the last few millennia can destabilize very quickly with small input changes

"DOUBLING" DESERVES ATTENTION

A *doubling sequence* has astounding effects and is key to understanding where SLR is headed. Doubling is a form of exponential growth and defies intuitive appreciation. A good illustration of the incredible growth is the challenge question of starting with a penny on the first day of a month and doubling it every day: two, four, eight, sixteen, thirty-two pennies, etc. On the thirtieth day, the penny would become more than five million dollars—$5,368,709, to be precise. In a thirty-one-day month, it would be twice that. Such is the power of doubling, a form of low-level exponential growth. When this is just a "bar bet" or trivia question, it's interesting, even astounding. As we will see in a moment, there is now some evidence that SLR may be following a doubling time progression, as has happened in the distant past.

Another illustration of doubling that's more relevant to rising sea level is the time it would take to fill a sports stadium with water, starting with a single drop, doubling the amount of water every minute: one drop, two, four, eight, sixteen drops, etc. How long do you think it would take for the entire stadium to be filled? Take a guess, then check the answer at the end of this chapter.*

The point is that the recent sea level stability or slow increase has lulled us into believing that it cannot rise catastrophically. That belief is without basis. Catastrophic SLR happened ten thousand years ago, even without our human-induced hyper-heating. We need to get a lot smarter if we want to have any possibility to be prepared.

> To provide some perspective on the challenge to reduce the warming, eighty percent of all of the energy generated in the US and globally today is from burning fossil fuels (coal, oil, and natural gas), which all produce carbon dioxide. Three countries—China, the US, and India—produce fifty percent of that carbon dioxide. Global carbon dioxide generation is increasing at 2.7 percent annually.

HOW MUCH SHOULD WE PLAN FOR?

Recent official estimates for SLR this century now have upper bounds as high as 8 to 10 feet (2.5 to 3 meters) with only minor contributions from Greenland and Antarctica.[9] The estimates have been continually raised over the last few decades, with the increasing evidence from the polar regions that the melt rate of glaciers and ice sheets is accelerating. Even half that amount—5 feet of global average sea level rise—would have the most profound, almost unimaginable impacts on people everywhere. Coastlines will change permanently, submerging vast areas, with geopolitical, economic, and humanitarian consequences. Many glaciologists now believe that to prevent SLR from exceeding 10 feet (3 meters), we have to get extremely aggressive about slowing the warming, which we are not yet doing.

We would also have to be very lucky, hoping that no catastrophic ice sheet collapses happen abruptly in Greenland or Antarctica during the next eight decades. Unfortunately I don't think we can count on that good luck. As covered in the last chapter, recently there are ominous signals coming from the polar regions. Even cautious glaciologists have been alarmed by the changes in the last three decades.

My present view is that responsible cautious planning for SLR this century should now assume 10 feet (3 meters). That's essentially double what I stated in *High Tide on Main Street* in 2012. The substantial increase is due to the continued rise of greenhouse gas emissions, the continued warming, and the increasing evidence of ice sheet destabilization in both Greenland and Antarctica. We should note that the year 2100 is just a commonly used benchmark for comparison. In the twenty-second century, sea level will continue to rise and will likely accelerate in the absence of extreme measures to slow the warming that is melting the ice sheets.

Konrad "Koni" Steffen was one of the most respected glaciologists studying Greenland and someone with whom I stayed in contact. In August 2020 he died at his "Swiss Camp" research station atop the Greenland Ice Sheet in a tragic accident. He fell into a crevasse, almost certainly caused by the unusual melting from the warming Arctic. In 2010, when I met Koni while researching my first book, I could not get him to go beyond a meter of sea level rise this century as a worst-case scenario. As a good scientist his positions were carefully reasoned. With that context, what he said in a powerful four-minute video from 2018 is all the more striking. He talks about five meters (16 feet) of SLR, possibly in fifty to one hundred years. That's an incredible change in his position in just a few years. To view the video: https://vimeo .com/306055715

For most of us, a much shorter time frame is going to be more useful. If we look thirty years out—within many people's personal planning horizon and even within the timescale of mortgages and other financial instruments—I estimate that we should be planning for at least a foot or two higher sea level, about half a meter. While that may sound modest, even that will have a profound impact on large coastal cities and small communities around the globe. Consider the heightened rate and impact of flooding due to just the few inches (centimeters) of higher sea level in the last three decades. The areas that routinely flood during king tides are already far larger than they used to be. Another six inches will greatly expand the flood zones, likely overwhelming drainage systems and causing more saltwater to intrude into the groundwater. Water and wastewater processing, underground electric cables, and telecommunication cables could all be affected.

Our sense of what is realistically possible is largely based on what has happened in the known and familiar past. "Black swan event" is now a term of art, from the eponymous 2007 book by Nassim Nicholas Taleb, describing high impact, rare, hard-to-predict events. The black swan theory and the "recency effect" cited in the foreword point to our instinctive belief that tomorrow will be like yesterday. That will not be true in many ways, particularly with climate. Rising seas and shifting shorelines are one of the most profound proofs that we are entering a new era.

Traditionally the worst storm, rainfall, or flood event of the last hundred years was deemed to define a "one-hundred-year flood event." From that it was deduced that the chance of such an event happening in the next year was one percent, one in a hundred. Such concepts are logical when we are in a *stationary* period, meaning that the forces and environment are generally stable. With the greater heat resulting from the greenhouse gases, the increased rainfall due to a warmer ocean, and new weather patterns as the Arctic sea ice disappears at an accelerating pace, we are definitely not in a stable period.

Though the melting Arctic sea ice has no direct effect on sea level, as covered in the previous chapter, it has three powerful effects on global warming, perhaps better thought of as **climate destabilization**: (1) As the large Arctic Ocean goes from bright white to very dark blue, it absorbs even more heat, similar to the way a white roof keeps a house cooler. (2) The disappearance of the thick ice that has covered the North Pole region for millions of years is one of the primary factors in the unusual weather patterns like the "polar vortex," as ocean and atmosphere now exchange heat differently without the 10 feet of year-round ice acting as a barrier to heat transfer. (3) Arctic thawing is releasing vast quantities of carbon dioxide and methane from the permafrost, adding to the greenhouse gases warming the planet, one of the **feedback loops** accelerating the entire process.

Our concept for preparedness generally assumes that the typical variations and extreme events will average out and may even follow a normal distribution curve. We typically apply this kind of thinking to the statistical probability of wildfires, deluge rain or drought, earthquakes, car accidents, and life expectancy. Until recently, it was reasonable to predict the chance of something based on the normal distribution of a large sample size of previous events. Entire professions, including insurance, engineering, and finance, were largely based on extending forward the last hundred years, or even thirty years. Such projections are valid, until the underlying forces change, which is what we have witnessed around the world in recent years. It is now common to get so-called *hundred-year* flood events from extraordinary rainfall every few years, and sometimes twice in one year. We see something similar with the extreme wildfires that are worse as temperatures warm, forests are in drought, and civilization brings humans and high-voltage electricity into the woods. Rain gets more intense as the oceans warm and evaporate more moisture up into the air. Even earthquakes, thought to be outside of human influence, became more frequent with the tremendous boom in fracking operations. Other examples of changes to baseline conditions and probability might include the chance of having a car accident, which changes with traffic congestion, driving speeds, and engineering advances. Human life expectancy changes with public health, medical technology, and lifestyle changes. In other words, when fundamental conditions change, so too will the frequency, patterns, and magnitude of related phenomena. Outdated concepts and constructs like hundred-year events and permanent shorelines can distort our awareness and perception of tipping points as we transition into a new era.

THE SURPRISING TRUTH ABOUT SEA LEVEL PREDICTIONS

Sea level is most directly related to the size of the glaciers and ice sheets, which are changing dramatically in real time. Each year we are seeing changes in the Arctic and in Antarctica that are new in the human experience. The term *glacial change* has always been used to suggest an imperceptibly slow process. Now glaciers are changing so fast that scientists are scrambling to keep up, making them even more cautious about predicting how much change to expect in the decades ahead. The accelerating trend of shrinking glaciers all over the world is unmistakable.

We all would like to know solid facts about our future so that we can plan. Vast efforts and expense are spent trying to see the future. But in most cases, when reality does not yield certainty, we routinely proceed regardless. Everyone plans for themselves and their loved ones, accepting the unknowns of their health, fate, and lifespan. Investors try to hedge bets against surprises; engineers and financiers build in margins of safety for the unexpected. Residents of San Francisco and Silicon Valley are smart and sophisticated, yet trade the very real possibility of a sudden, massive earthquake for the compelling appeal of the San Francisco Bay Area. As exemplified by astronauts, pilots, race car drivers, police, firefighters, soldiers, and other specialists, we can assess risky situations and then put the worrisome risk out of our minds to proceed with the task at hand. We must learn to do that with the "known unknown" of the rising sea.

To meet this challenge, and to do it smartly, we must understand why the rate of rise is unpredictable, as well as why it can be slowed, but not stopped, for centuries. From those two realizations comes the urgency to begin adaptation sooner, rather than later. The chart below shows the wide range of projected SLR, as predicted by the official US National Climate Assessment in 2018. It may look clear and simple, but it displays a stunning fact that few want to recognize. Note the multiple lines fanning out to the right and the extremely wide spread, showing 1 to 8 feet as the range of where sea level may be in 2100. That huge spread is a good indicator of our *inability* to know the future. Also, the projection lines are smooth curves. As shown in the preceding graphs, sea level has not risen smoothly in the past and will almost certainly not do so in the future. Taking the average or middle line may seem safe but makes as much sense as planning for a medium hurricane or earthquake. It would be much smarter to use the highest line and hopefully have a margin of safety.

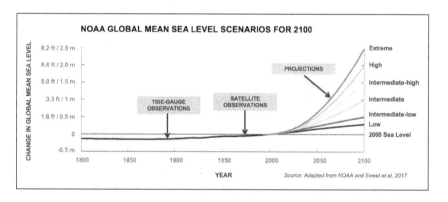

Figure 6. Sea level rise cannot be predicted with any certainty. This official 2017 US government graphic shows a wide range of curving lines for possible sea level rise, demonstrating the difficulty of predicting SLR in future decades.[10] Also, it will not likely follow any smooth curved line. In the past the rate of sea level rise has changed abruptly, and it is likely to do so in the future, particularly with the current changing rates of ice melting on Greenland and Antarctica.

The wide variation of projected SLR might suggest poor science, but that is not the case at all. The fanning out of those lines is essentially due to two things. First, no one knows how warm the planet will actually be by the end of the century since we don't know the total energy demand, nor whether it will be produced by fossil fuels, nuclear, or renewables. Second, as we pointed out in Chapter Two, the collapse of the mile-high ice sheets on Greenland and Antarctica will be chaotic, defying precise prediction. One other reason for confusion about different "predictions" for future sea level is a basic misunderstanding about models.

PROJECTIONS, NOT PREDICTIONS; THEY'RE JUST MODELS

Of course, all manner of things are modeled, from labor and economic metrics, to flu deaths, to the growth of investments, to our lifespans. They are understood to be estimates and often range widely. Anyone who has bought life insurance or developed retirement investment plans likely has seen a projection for their lifespan based on several factors, such as their parents' lifespans, their body mass, whether they smoke, etc. For example, a thirty-year-old person might be given a projected lifespan of seventy-six years. While it's perhaps a useful guideline, most people understand they should plan for a much wider range of possibilities for their longevity. Common sense tells us that the model's *projection* for how long we might live is only a tool to give us some idea but is not a *prediction* whatsoever. Sea level projections and models should be looked at with the same perspective.

Furthermore, different professions that need to deal with future flooding have very different needs and perspectives, causing them to use different models. Consider three different views of water-level projections for scientists, engineers, and those involved with emergency management:

- *Scientists* strive to understand all the factors, usually looking out to an agreed reference year of 2100, stating what they can explain will happen, quite specifically, with high statistical confidence. They have to defend their views to a jury of peers based upon rigorous rules, which warrants a cautious, conservative approach. New concerns and "possibilities" are often redirected for further study and research or covered in an obscure footnote that few will read. Accordingly, scientific estimates can be quite cautious on the low side. The finest example of the scientific community's work is the UN Intergovernmental Panel on Climate Change, referred to as the IPCC. Though excellent overall, unfortunately the IPCC has a weakness regarding projected SLR that I call the "Antarctic Asterisk," to be explained shortly.

The sixth edition of the IPCC Assessment Report is scheduled to be issued in 2022; updated information is available at www.ipcc.ch. The projections for future sea level, and my assessment of their shortcomings in the current version, the fifth edition (2013), are covered later in this chapter.

- *Engineers* designing structures to be flood tolerant will be cautious in a very different way from scientific caution. They want to design for something approaching the worst possible situation, often using one hundred years as the design life. They want a factor of safety, particularly where there is uncertainty about how high the water will reach. To be safe, they would very likely design with a more extreme high-water mark in mind than the scientific figure.
- *Emergency management officials and flood forecasters* have a very different view than those of scientists or engineers. They are focused on the near-term possibilities that can combine unexpectedly. Their job is to be able to respond to whatever comes, with a priority to save lives and property and to support recovery. In planning exercises, they will consider worst-case scenarios regardless

of whether they meet the requirements of peer-reviewed science journals. They have to do *triage*, making tough decisions on the spot, often with limited information. They want the latest, most comprehensive models to give flood warnings on a dynamic basis. While they will work with scientists and engineers closely, their mission is public safety. To state it simply, when flooding is imminent, they need to know where to put out the sandbags, how high to pile them, and when to call for an evacuation. In the United States, FEMA, NOAA, and the Coast Guard are prime examples of agencies with the emergency management perspective.

To compare the very different estimates for rising sea level used by scientists, engineers, and emergency managers, see the online Deeper Dive Note #4, **Sea level rise projections—wide range, e.g., Florida**. It compares the estimates from the IPCC and NOAA. Just in the next twenty years, the estimate varies from 10 to 21 inches, with far larger spreads at fifty and one hundred years. Though these projections should cause deep concern for anyone in southeast Florida, the message is similar for thousands of coastal communities.

Models for future sea level and flooding are very different for each of those three sectors. In later chapters I will cover some other professional communities who will have additional perspectives and attitudes about planning for future flooding and SLR, including property owners, insurers, and financers. For now, it's just important to recognize that there are valid reasons for the great diversity of future sea level models and estimates.

Sea Level Will Surge Upward in 2025. In 432 BC, the astronomer Meton of Athens identified an 18.6-year "lunar nodal cycle," often rounded up and described as a nineteen-year cycle. That pattern, the Metonic cycle, is due to orbital patterns and the gravitational forces of our Moon and Sun. It's a repeating up-and-down two and a half inches (6 cm) that adds to and subtracts from high tides. By itself, it's not a huge amount, but it's greater than the present annual amount of SLR. Thus it has the effect of essentially doubling—or canceling out—the apparent annual SLR for the better part of a decade. This phenomenon adds to the confusion about SLR. From 2015 to 2023, the lunar nodal cycle is in the "down phase," tending to hide global SLR. In 2025 it will be back in the "up phase," appearing to greatly increase SLR. (For a diagram and more information, including the relation with Greek, Hebrew, and early Chinese calendars, see **19-Year Lunar Cycle** in the online Deeper Dive Note #5.)

Perhaps the biggest source of confusion about future sea level stems from a misunderstanding about the huge, comprehensive IPCC report, generally considered to be the definitive authority on climate change. Updated every five or six years, this excellent scientific study is compiled by a global group of more than two thousand experts. They operate very transparently, publishing their draft versions and answering every question and suggested revision online with full accountability to the source of information. In the most recent IPCC report, the projection for SLR is likely the single most controversial and significant item in its thousand-plus pages. At its most fundamental level, the problem is due to a protocol or methodology. From the IPCC's start in 1988, the carefully stated policy was effectively *only to include facts that had been peer-reviewed in a professional journal, that were represented as objective values, with a high degree of certainty, with the year 2100 as the benchmark.*

It was a very good process in principle and in general application. The only real exception is how to project sea level, given that the collapse and melting of glaciers and ice sheets cannot be specifically quantified to such a high degree of certainty. In effect, the IPCC cannot come up with a realistic number that satisfies their own rules described above. Essentially, they are not asking what *could realistically* happen, but rather asking what *will happen*, expressed with a high degree of certainty. Those are two very different questions. Most people would like to know the answer to the first.

"ANTARCTIC ASTERISK" HIDES SLR REALITY

It even surprises many experts to learn how severely the IPCC understates projected sea level rise. Paradoxically, the omission is due to their good science and rigorous discipline. Since they can't agree on exactly how much Antarctic melting will contribute to ocean volume this century, they have effectively used an asterisk. The uncertainty is covered in a footnote to the huge report that few read. Many scientists agree with my belief that the report presents the case for rising sea level poorly. As I will elaborate in Chapter Seven, even highly respected research companies and government organizations around the world often look only to the IPCC's top-line figure of less than one meter (about 3 feet) as the worst-case scenario for future sea level this century. Often they don't notice that Antarctica—which holds *eighty-eight percent* of potential SLR—is largely omitted from the projections. While I appreciate their dilemma and their wish to follow protocol, it is dangerously misleading.

To be specific, the IPCC assessment runs four scenarios for climate change, ranging from a "best case" with drastic reductions in greenhouse gases and global warming (RCP 2.6) to the "worst case," business-as-usual, largely following our current path of growth, consumption, and increase in greenhouse gases and temperature (RCP 8.5). In that upper-end scenario, their top-line figure for global sea level rise by the end of this century is 82 centimeters, or 32 inches. That figure is the most commonly referenced figure in the world for future sea level. In their worst-case scenario, with global average temperature warming almost 9° Fahrenheit (5° Celsius) by 2100, it includes just 6 inches (15 cm) of sea level rise from Antarctica. As mentioned in the last chapter, that's out of a potential 186 feet of sea level contribution from the southern continent. That 6 inches is not at all representative of what the glaciologists believe to be at stake. Think of it more as a *placeholder* until we have better information. I call it the "Antarctic Asterisk."

There are other projections for worst-case sea level rise, higher than the IPCC's. Recent studies from 2017 to 2019 by the US government, the British government, California, and New York State are all now looking at high-end scenarios of 8 to 10 feet (2.5 to 3 meters) for SLR this century. They are able to change their protocols more easily than the huge international IPCC. Still, they are unable to answer the question of the probability of such a worst case. That comes back to the unpredictability of the rate of warming in the coming decades and the mysteries of glacial collapse.

"Antarctic Asterisk"—The IPCC Report and the understated sea level projections are explained in Deeper Dive Note #6 in the online material at www.movingtohigherground.com. The previously cited Deeper Dive Note #2 gives more information about Antarctica, and possibly more up-to-date information.

Unfortunately, it's not realistic to wait to adapt to rising sea level until we know precisely how high and how soon it will rise. As with anticipating earthquakes or pandemics, we need to plan for rising sea level and flooding *without* knowing the specific timing or the magnitude. While that may seem overly cautious, we fully accept unpredictability when it comes to planning for other risks. For example, no one doubts there will be another severe earthquake in the San Francisco area, perhaps even worse than the famous 1906 quake rated to have been a 7.9 magnitude. Yet we have no idea when the next one will occur. Even with hundreds of strain gauges now monitoring the fault lines, the guidance for building codes and construction is essentially: There is a ten percent chance of a magnitude 8.0 or greater seismic event in the next fifty years in the Bay Area. Could that be more vague?

Yet that's the basis for their planning, engineering, architecture, and emergency response. We plan and design for another big earthquake in San Francisco and in dozens of other places around the world where there are active fault lines and zones of tectonic plate convergence, despite having no idea when it will occur or what its magnitude will be.

Making the challenge of planning for SLR even harder is the fact that the estimates vary hugely. Let's take a closer look at why it is difficult to project the rate of SLR from melting ice sheets. Visualize an ice cube on a table. How many seconds it will take to melt depends on many factors: temperature in the room, air currents, the closeness of your body heat and exhalation, and the density of the ice—which varies with the pressure where it was formed. Size also matters, though even ice cubes from the same freezer tray will often melt at slightly different speeds. Predicting the potential rate of melting of the two gargantuan ice sheets and glaciers is both similar *and* far more complicated, given that we are talking about ice sheets the size of North America, which are miles thick, working their way over the rough, mountainous terrain of Alaska, Greenland, and Antarctica. When ice is more than 150 feet thick, it deforms rather like modeling clay. Referred to as plasticity, that bending and reshaping greatly affect the time it takes for the gigantic glaciers to reach the sea, breaking off as new icebergs or becoming rivers of meltwater, the two primary causes of future SLR.

BETTER MODELS

One tool that can help us understand, visualize, plan, and adapt for extreme changes is better coastal flooding models. They are one place where risk gets translated into practical effect. Great improvements have been made with models in recent years, but there are significant limitations. At the risk of oversimplifying, we should distinguish two entirely different types of models:

1. **Numerical climate models** take into account thousands of variables, coupling the effect in the oceans and atmosphere to see how our entire climate changes, both in the short term and in the longer term. Even where they may

be focused on a local area for impact, they must have a global base to capture the complexity of the various factors' interactions. These are among the most demanding computer applications of all and are run on supercomputers or their equivalent in the cloud. The output of these numerical global models may be shown in visualizations, somewhat similar to the kinds of maps we are all familiar seeing with weather forecasts.

2. **Flood visualization models** work with the relevant variables of sea level, storms, rain, tides, current, and local topography. Several free flood models are available for the public and businesses to assess flood vulnerability, possibly down to the parcel level. Three that are based in the United States are:

- Flood Factor—www.floodfactor.com
- Surging Seas/Climate Central—https://sealevel.climatecentral.org/
- NOAA's Digital Coast—https://coast.noaa.gov/digitalcoast/

The first two are offered by nonprofit organizations, and the third is a US government product. Flood Factor is the most sophisticated in terms of combining storm surge with rising sea level. Each operates somewhat differently in terms of available coverage areas and in how far out in the century they look. Such products are frequently being improved, particularly in terms of their user interfaces.

In addition to those products for flood risk visualization, there are also some private-sector approaches to climate modeling that include flooding. Jupiter Intelligence is perhaps the highest profile of the few firms in the growing field. Coming out of the culture of Silicon Valley, they have pioneered some innovative ways to take climate modeling and related risk prediction to the next level. Their products for flooding, temperatures, and wildfires promise to improve the specifics of time and location by at least an order of magnitude (a factor of ten). By using private capital, having the freedom to innovate, and developing products based on market demand, they aim to transform the field. Given what's at stake, we should be glad that government, academia, and the private sector are all working to improve the suite of tools generally described as models.

HOW MANY FEET PER DEGREE OF WARMING?

Perhaps the key question is: How high will sea level rise for each degree that the planet warms? Though there are some diverse estimates in scientific journals,[11] a rough answer is "hiding in plain sight" in that 400,000-year chart in the first chapter. The repeating pattern of vertical sea level change is close to 400 feet, or about 120 meters. The average temperature swing for the same periods of time is 9° Fahrenheit (5° Celsius). That relationship, based on facts of physics as simple as the temperature at which ice melts, holds a profound and potent message that we ignore at our peril.

Our planet has already warmed almost 2° Fahrenheit (approximately 1° Celsius) in the last century or two, the era of burning increasing quantities of coal and petroleum. With the excess heat already stored in the oceans, and the historical relationship between sea level and global temperature, the science is crystal clear. Though it could take centuries for the ice sheets and glaciers to melt, when they reach the new equilibrium with the planet at its current temperature, global sea level will be roughly 70 feet (~20 meters) above present, changing the shape of all the continents and islands.

We must face this simple, indisputable truth. Sea level will rise for centuries. Even without any further warming, we are headed to the situation that last occurred one hundred twenty thousand years ago, when sea level was twenty five feet higher than today (approximately 7 meters). Even the 10 foot rise that is now realistic this century will be catastrophic and will change the world as we know it. Regardless of whether that amount occurs in fifty years or a hundred, now is the time to start aggressive adaptation.

While we don't know how many centuries it will take for that to happen, we do know that the faster the planet warms, and the faster the ice melts, the faster ocean heights will rise. Even if the temperature goal of the 2015 Paris Agreement, or a similar plan, can somehow be achieved, the continued warming means faster ice melting and higher sea level than we are presently experiencing.

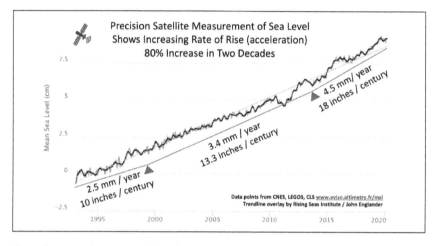

Figure 7. Starting in 1992, satellite altimetry has yielded much more precise sea level data. The squiggly line is the actual data points. The three trendline segments show that the rate has increased from 2.5, to 3.4, to 4.5 mm a year, a significant increase in slope. In inches those rates would mean an increase from 10 inches, to 13.3 inches, to 18 inches in a century. It is this rapid increase in the rate of SLR that has scientists concerned.

Generally in this book, SLR (sea level rise) refers to the change of global *average* sea level, scientifically often called global mean sea level (GMSL). Over the twentieth century, the average rate of SLR was less than 2 millimeters (mm) a year, about seven hundredths of an inch. Though the amounts are small, it's the acceleration and trend that should get our attention. As shown in the chart above, starting in 1993, with the inception of precision satellite measurement of SLR, the rate of rise was 3.1 mm a year in just the first two decades, a fifty percent increase from the 1.9 mm average rate in the twentieth century. Notice that in the most recent decade, 2010–19, the rate of SLR is 4.5 mm a year—another fifty percent increase. That's almost two-tenths of an inch a year, quite an increase from seven hundredths of an inch. This suggests that we may already be in doubling mode. The next decade should give a much better idea of the rate of acceleration. If we go back to the sports stadium example, we get a sense of where this could be headed, despite the fact that we don't know how hot the world will actually become because of the uncertainty about energy policy and greenhouse gas levels. So, we can't know how fast sea level will actually rise. But as a final example of exponential growth, let's just follow this hypothetical sequence with doubling every decade. For simplicity, we start with a quarter of an inch a year, about 6 mm, of annual sea level rise, close to the last figure on the satellite date chart above.

Hypothetical Sea Level Rise per Year Based on Doubling Every Decade

Year	Inches Per Year (approx. metric)	Per Decade Equivalent
2020	0.25 (~6 mm)	2.5 inches (~6 cm)
2030	0.5 (~12 mm)	5 inches (~13 cm)
2040	1 (~2.4 cm)	10 inches (25.4 cm)
2050	2 (~4.8 cm)	20 inches (~50.8 cm)
2060	4 (~10 cm)	3 ft, 4 in (~1 meter)
2070	8 (~20 cm)	6 ft, 8 in (~2 meters)

To be clear, this is just a mathematical exercise to show doubling rates or exponential growth. There is absolutely no way to know now if SLR will reach a 4-inch (nearly 10-cm) *annual* rate of increase by the year 2060. But, if it did, that would mean more than 3 feet (1 meter) of higher water level in a single decade. Most oceanfront communities and infrastructure would find it extremely difficult to cope with that rate of increase, just for a single decade.

Accelerating Rate of Increase. It is easy to miss the differences among (a) a one-time, fixed amount increase to sea level; (b) an increasing rate of SLR; and (c) an accelerating rate of increase. Think of the differences with water flowing into a bathtub and out the drain. If the two flow rates are the same, the water level in the tub stays constant. If you open the faucet a little more, the water level will rise over time, but at a constant rate. If you keep opening the valve more and more, the **rate** of rise would get faster. The latter is what is happening now with the rate of ice melting, particularly in Greenland, causing global sea level to keep rising faster and faster. As the planet is getting warmer and warmer and the ice is melting faster and faster, the rate of sea level rise is getting faster each decade.

The thesis of this book is that the only smart thing to do is to adapt for higher sea level, sooner rather than later. And it's worth noting—from Figure 5 in Chapter Two, showing the 400-foot rise since the recent ice age—that sea level has risen at extreme rates in the past, even prior to the extreme rate of warming at present.

To recap, global average sea level is rising from two effects: (1) the melting ice on land, and (2) the fact that warmer oceans expand slightly ("thermal expansion of seawater"). As long as we keep applying increasing amounts of external heat to the Earth system, the rates of melting ice and rising sea will grow and accelerate until the ice is gone.

US DEFENSE DEPARTMENT: TIME FOR EXTREME SCENARIOS

With such uncertainty, huge risk, and complex dynamics, and with global and national security at stake, it should not be surprising that military science has some valuable perspective. In my opinion, the best assessment of the risk of sea level rise to date is the 2016 study led by the US Department of Defense, *Regional Sea Level Scenarios for Coastal Risk Management: Managing the Uncertainty of Future Sea Level Change and Extreme Water Levels for Department of Defense Coastal Sites Worldwide.* Their evaluation concluded that SLR this century could be 2 meters (7 feet) or more. More important was their emphatic point that we needed to begin planning based on extreme scenarios, rather than expecting to have a precise number. This statement from the report's Executive Summary says it well:

> *The decision-making paradigm must shift from a pre-dict-then-act approach to a scenario-based approach. As a decision-maker, the fallacy and danger of accepting*

a single answer to the question "What future scenario should I use to plan for sea level change?" cannot be stressed enough. Those used to making decisions based on a "most likely" future may have trouble relating to this reality; however, a variety of uncertainties, including the uncertainties associated with human behaviors (i.e., emissions futures), limit the predictive capabilities of climate-related sciences. Therefore, although climate change is inevitable and, in some instances, highly directional, no single answer regarding the magnitude of future change predominates. Traditional "predict then act" approaches are inadequate to meet this challenge.

There are several reasons why we cannot accurately forecast future sea level in the coming decades.

1. Major geophysical events do not lend themselves to specific predictive modeling.
2. The last time the ice sheets and glaciers collapsed was over ten thousand years ago. Even looking back at those known events, we cannot interpret the geologic record about the rate of rise per year or even decade. Geology usually looks at centuries or longer.
3. Today the rate of warming is hundreds of times faster than in millions of years, causing a different rate of melting ice and rising seas. If you double the amount of warming, you do not immediately get double the rate of ice melting. There are some nonlinear aspects due to heat transfer (thermodynamics), crystalline structure of the ice, changing ocean currents, etc.
4. Our society is still arguing about energy policy—questions such as the continued use of coal, tar sands, and natural gas, and economic incentives to switch to renewable sources. Without knowing how we will make our energy in the coming decades, it is not possible to know the amount of heat that will be added to the Earth system,

which ultimately is what changes the size of the ice
sheets, changing global sea level.

In short, it is not possible to accurately predict future glacial collapse and sea level rise down to inches per decade or mm per year.
This 2016 assessment, led by the Department of Defense but involving
multiple agencies, clearly concludes that we cannot afford to wait until
we know the numbers with certainty. We must begin the hard work of
designing and building for the future if we want to limit the devastation caused by the disruption to the global supply chain, financial markets, national security, mass migration, and humanitarian demands.

Looking at low-probability, high-risk scenarios is an important
part of military leadership training and strategic planning. With rising
sea level, the threat hits home for the military. From the world's largest
navy base, Naval Station Norfolk, to tiny Diego Garcia island in the
middle of the Indian Ocean, the military has to consider the long-term
strategic logistics for future missions. The US military has proven itself
exceptionally capable and adaptive. Adapting to more than a meter of
SLR will present a new set of challenges. An analysis in 2016 by the
Union of Concerned Scientists showed that 128 US defense facilities
are threatened by just 3 feet (1 meter) of higher sea level. In the United
States and elsewhere, the military is fully exposed to the SLR challenge. Their approach to risk and uncertainty may enable them to do
a better job of planning, preparation, and adaptation than the civilian
and commercial sectors.

---

* ANSWER to challenge question above about time to fill a soccer stadium:
Forty-seven minutes, give or take one minute, depending on the size of the
stadium.

4

REDUCING CO$_2$ AND BEING RESILIENT ARE NOT ENOUGH

*Whenever a new and startling fact is brought
to light in science, people first say, 'it is not
true,' then that 'it is contrary to religion,' and
lastly, 'that everybody knew it before.'*
—Louis Agassiz, famed geologist

Confusing the drive for clean air with fighting climate change is common but makes it impossible to have an intelligent discussion about solving the problem that is causing the sea to rise. Surprisingly, the scientific and environmental communities are largely to blame. They routinely talk about adding or removing tons of *carbon* to or from the atmosphere as a shorthand when they really mean carbon dioxide. No wonder that some political leaders have famously mixed up climate change with the issue of having clean air. Before we look at solutions to the rising sea, it's essential to clear up the confusion.

The chemical compound carbon dioxide (CO$_2$), the greenhouse gas that's largely causing the warming, is completely invisible (clear). The chemical element carbon, by contrast, is pure black. In the air carbon can be seen as dark smoke or a hazy smog that limits visibility. When it falls to the ground, it's usually referred to as *soot*. Confusion about the basics of black carbon versus clear carbon dioxide is widespread and an obstacle to intelligent discussion. Calling both carbon and carbon dioxide "air pollution" as is now common, adds to the problem.

In fact, three very different issues are commonly mingled in error, even by those concerned about climate change. Carbon, carbon dioxide, and the *hole in the ozone* are not at all similar and require different responses. First, we should reduce carbon pollution in the air to maintain good air quality. Second, we must reduce carbon dioxide emissions to slow and eventually stop planetary warming. As a very distinct third issue, to avoid ultraviolet damage, we should continue the 1987 protocol to reduce certain refrigerant gases, to repair the hole in the ozone. They are three entirely separate issues. Of the three, only carbon dioxide is significantly responsible for rising temperatures, melting ice, and sea level rise. This is not some nitpicky semantic point. Clear communication is critical if we are to prioritize policies, programs, and budgets. Confused descriptions lead to totally ineffective policy prescriptions. A great example is this May 6, 2017, headline, from the website of WLRN, the Miami NPR station. It is completely wrong and misleading in several respects:

> *How to Prepare for Sea Level Rise: Miami Developer*
> *Recommends Cutting Carbon Emissions*

Technically, it should say *carbon dioxide* instead of *carbon* to make the distinction between particulate carbon in the air and clear greenhouse gases. But even changing it to "carbon dioxide emissions" would not make this headline true. If a building—or even, for that matter, the entire city of Miami—reduced CO_2 emissions to zero, that would have no effect on its exposure to rising sea level. CO_2 emissions blend globally, high up in the atmosphere, slowly warming the atmosphere and oceans. Over the course of many *decades and centuries*, that warming affects the amount of ice on land, which in turn determines global sea level.

The opening of that headline, *to prepare for sea level rise*, is completely misleading. Preparing for higher sea level and more short-term flood events might require a protective barrier, sealing some low openings, installing pumps, elevating the building, or relocating, but not reducing carbon dioxide emissions. What made this erroneous headline especially disturbing is that the article was reporting on an all-day conference in Miami for professionals about sea level rise. If any group should have been aware of the difference between reducing greenhouse gas emissions and preparing for sea level rise, it should have been clear in the discussions at that meeting. To be fair, as soon as I contacted the NPR station and pointed out the fallacy, they changed the headline.

It's a good example of how incorrect or misleading phrasing can perpetuate totally incorrect perceptions of problems and solutions. To solve a problem, we must understand it. To understand a problem, we have to use language, concepts, and *memes* that are both true *and* intuitive, so that they can be comprehended by broad audiences. That's why I try to avoid all jargon and acronyms and strive to use plain language and imagery. That conference and headline are not an isolated case. It is common for people to confuse the reduction of greenhouse gases and warming (mitigation) with the very real challenge of dealing with increased flooding from the rising sea (flood hazard mitigation).

ALL THINGS "GREEN" WILL NOT SOLVE THE WARMING

Regarding the topic of communication, some ardent environmentalists equate everything "green" and environmental as being not only good, but mutually consistent. For example, many people assume that more recycling or the elimination of plastic bags can help solve SLR. To be clear, I very strongly support both causes, recycling and ending the deadly scourge of plastics in the ocean that is a blight on our ocean planet and kills marine life. They are both terrible problems that need to be addressed. Yet the truth is that the melting of the ice sheets and glaciers is quite unrelated to those two problems. We could recycle everything possible and eliminate all present and future plastics in the sea without any noticeable effect on the planet's average temperature, the rate of ice melting, and sea level rising. We can and must work on several different issues simultaneously.

Putting all environmental issues in a virtual "big green bucket" is misleading and can be ineffective. If we are going to get smart and really tackle the big problems like sea level rise, climate change, plastic pollution, lack of fresh water, impacts on wildlife, humanitarian concerns, and wasteful use of materials, we need to be accurate about which action will solve which problem, accomplishing which goal. Sustainability and resiliency are very popular words these days, and each does have some relevance to sea level rise and flooding. But even those terms can convey incorrect messages about rising sea level and the greater frequency of flooding from other causes. While the two terms have wide-ranging use and connotation, this is how I use them for clarity, in contrast to *adaptation*:

- *Sustainability* applies to products and processes that can be used continuously without depletion. Renewable energy such as solar and wind and the recycling of bottles or aluminum cans are good examples of sustainable practices that conserve resources and reduce waste. As noted above, recycling can have important benefits but is not a way to slow or stop climate change and rising sea level.
- *Resiliency* applies to the ability to defend, to maintain operations, or to recover after a disruption. For example, better design could reduce flooding potential during a storm and also could facilitate resuming operations more quickly following an interruption. In the next few chapters, we will consider some design concepts relevant to resiliency.
- *Adaptation* to rising sea level is quite different from resiliency and is the primary focus of this book. SLR is essentially *permanent* flooding or submergence that warrants different approaches than dealing with short-term flooding in a particular location.

Rising sea level is not a disruptive "event" but rather a slow change moving the baseline higher. The rising ocean level will not recede, allowing recovery, as happens with all the other forms of flooding. Furthermore, higher base sea level will raise the water height of all future coastal flood events that much higher as well.

CO$_2$ AND TEMPERATURE: WHICH FORCES WHICH?

The current focus of concern about climate change is that rising CO$_2$ (carbon dioxide) levels associated with fossil fuel use are pushing temperatures higher. But long before the burning of fossil fuels, such as back in the ice ages, temperatures also changed. Understandably, many skeptics say that would appear to contradict the concept that higher CO$_2$ is causing global warming. That underlies a lot of the confusion about human-triggered warming and needs to be explained. Does higher CO$_2$ cause temperature to rise, or does warmer temperature cause CO$_2$ levels to rise? Surprisingly, it turns out that both are true. Global carbon dioxide and temperature levels move in synchronization due to two different but simple reasons. Either one can lead and the other will follow, over a period of centuries. This paradox is actually easy to explain:

- CO$_2$ follows temperature. In the last chapters we looked at the ice age cycles, the natural, roughly hundred-thousand-year pattern of heating and cooling, triggered by Earth's changing orbit, that has dominated climate change for over two million years. When the oceans are in a natural warming phase, they release gases, including carbon dioxide. A simple demonstration will show the underlying physics. Take two bottles of cold carbonated beverage. Uncap them. Warm one bottle. It will go "flat" much quicker, demonstrating that warmer liquids hold less dissolved gas. Thus, rising temperature releases carbon dioxide, causing higher levels in the atmosphere. The reverse is true as well, with more carbon dioxide being stored in the sea during cooling eras.
- Temperature follows CO$_2$. The term *greenhouse effect* comes from the common method to grow plants in cold temperatures. Sunlight passes through a glass roof. When it hits the ground, it turns into radiant energy, heat. The glass then acts as a heat barrier, keeping the warm air inside. CO$_2$ high up in the atmosphere works the same

as the glass, passing the sunlight and trapping heat, as demonstrated two centuries ago (see sidebar).

The fact that carbon dioxide (CO_2) traps heat in the atmosphere was proven in 1859, and first publicly demonstrated by Professor John Tyndall in London at the Royal Institution. The concept that CO_2 trapped heat and could affect global temperature had been understood due to the earlier work by the French mathematician and physicist Joseph Fourier in 1824. But Tyndall is the one who demonstrated and measured with precision what we now refer to as the **greenhouse effect.** The science that fossil fuels would cause warming was advanced in 1897 by Svante Arrhenius. The Swedish physicist/chemist calculated how much an increased level of carbon dioxide would actually increase global temperature. Despite the scientific technology being relatively primitive, he got the values quite close to what we now know. That's a testament to his brilliant analysis and the simplicity of the underlying science. This is the proof of **causation**, not just **correlation**, between carbon dioxide and the warming atmosphere.

Carbon dioxide has not been at the current level, above 400 parts per million (ppm), in at least three million years. When looked at in the scale of that first chart in the first chapter, the line is going straight up. It is this path of CO_2 and the correlation with global temperature that is at the center of concern. There is no way to quickly get CO_2 back to the historic range of 180 to 280 ppm.

To get the latest measurement of CO_2, showing the annual trend and seasonal cycling, go to http://www.esrl.noaa.gov/gmd/ccgg/trends/. The famous "Keeling Curve" showed there is undisputed proof that CO_2 is rising sharply. With simplicity and precision, it has been documenting global carbon dioxide since 1958. For more about **temperature, CO$_2$, and SLR correlation**, see online Deeper Dive Note #7, at www.movingtohigherground.com.

Reducing carbon dioxide emissions is extremely important but must not be confused with delaying or preparing for higher sea level. Based on the geologic record and basic physics, for sea level to start falling the ice sheets would have to "grow." For that to happen, global temperature would have to be several degrees colder. For that to happen, carbon dioxide needs to be dramatically reduced. At the current level of carbon dioxide and temperature, ice will continue to melt and the sea will rise. Cutting global carbon dioxide emissions is essential, but presently they are still increasing. I think it is possible that during this century we will find the technology to produce sufficient energy while reducing the level of atmospheric CO$_2$. That can eventually slow the warming, which would slow the rate of melting ice and rising seas somewhat. Reducing emissions at a single office building, city, or nation will not directly slow SLR. Even if the entire world could instantly switch to one hundred percent renewable energy, such as solar or wind, sea level will continue to rise due to the excess heat already stored in the ocean, which has an effect of melting the ice on land.

Because CO$_2$ has now shot upward like a rocket, temperature is following, but the two are far out of balance. The global level of greenhouse gases determines the insulation value of the atmosphere and the effect on temperature. Increased warming does affect the size of the glaciers and ice sheets, but the effect is very delayed. Reducing CO$_2$ emissions and concentration level in the atmosphere is of paramount importance despite the fact that it cannot soon stop SLR. If we do not work to slow the warming urgently, the melting and the rising sea will assuredly happen faster. Referring back to the primary graph from Chapter One, over long periods of time sea level tracks very closely with carbon dioxide and temperature. The last time CO$_2$ was at the present level, above 400 ppm, was several million years ago (in the

Pliocene). Back then, global average temperature was just a few degrees warmer than at present and sea level was approximately 80 feet above present (25 meters).[12] Coupled with the temperature analogy in the previous chapter, this is strong evidence that we must work to reduce carbon dioxide levels as global priority number one. At the same time, we need to anticipate much higher sea levels.

THE HOLE IN THE OZONE

The hole in the ozone layer is another issue that gets confused with carbon dioxide and carbon. The ozone layer high up in the atmosphere protects humans and wildlife from harmful ultraviolet radiation that can cause sunburns, cataracts, and skin cancer. By the 1980s, a growing hole in the ozone layer was identified. Research showed that refrigerants and the propellant gas in aerosol cans were major factors. Those chlorofluorocarbons (CFCs) were targeted. The Montreal Protocol, an international agreement, was signed in 1987, phasing out their use. It's a tremendous international success story. Often it's put forward as an example that we could do the same thing with carbon dioxide. There is a parallel, but finding a substitute for CFCs in aerosols and refrigeration was considerably easier than finding a way to eliminate greenhouse gases associated with all fossil fuel combustion. The companies that made the CFCs quickly set about to create substitutes that they could also sell, in many cases for even greater profits. The transition from fossil fuels to renewables is not quite as simple.

"Ocean acidification" is a rather obscure effect of CO_2 emissions that will have a truly enormous impact on marine ecosystems, particularly shellfish and corals. As we put more and more carbon dioxide into the atmosphere, nearly twenty-five percent of it dissolves into seawater, creating carbonic acid. This is lowering the pH of the oceans, making them more acidic. While not directly related to SLR, this is an important issue that also results from the higher level of carbon dioxide in the atmosphere. See online Deeper Dive Note #8 at www.movingtohigherground.com for more about ocean acidification.

Methane is a complex and extremely important issue related to global warming, which indirectly affects SLR. In its pure form, methane is more than a hundred times more potent than CO$_2$ as a greenhouse gas. Methane is rather unstable in the atmosphere. Over decades, the clear gas chemically transforms to carbon dioxide, which is extremely stable, with lasting effects on global temperature. With the increases in fracking and the use of natural gas, methane levels in the atmosphere have increased substantially. In addition, there are concerns about natural sources of methane, including cows, permafrost thawing, and releases from the seabed.

For more about **methane**, see online Deeper Dive Note #9 at www.movingtohigherground.com.

There are now a multitude of efforts all over the world aimed at reducing greenhouse gas emissions, particularly carbon dioxide. These range from the development of renewable energy sources, to greater energy efficiency in buildings and transportation, to battery technology enabling better use of intermittent renewable energy sources. While the goal is to slow the warming, it can't happen quickly enough to eliminate the threat of rising sea level. So we must tackle the problem from both sides: slowing the warming and adapting to the changes that are now unstoppable.

NO MAGIC SOLUTION TO STOP THE RISING SEA

Many people express hope that we can apply technology to "solve" rising sea level the way that we have done with other big challenges and leaps forward. Examples are frequently offered, such as putting a man on the moon, medical miracles, or the amazing world of electronics. Unfortunately, none of those great feats of engineering apply to the challenge and basic physics that a warmer planet will have less ice and higher ocean levels. No doubt we will continue to push boundaries of transportation, mechanical engineering, medicine, and electronics. Ice, however, will always melt at the same temperature. On a warming planet there will be less ice. As ice moves from the frozen form on land into the sea, the vast volumes of water will raise the height of the ocean, moving shorelines inland. "Solutions" need to focus on how to live with rising sea level, our focus in the next two parts of this book.

GEOENGINEERING

Beyond the important efforts to reduce the amount of new carbon dioxide that we are adding to the atmosphere, there are two areas of research and development that are considered climate engineering, or "geoengineering." These are purposeful efforts to reduce or counterbalance the warming effect of the greenhouse gas emissions. Such efforts are now being pursued in spite of many experts' grave concerns that they are risky and could do even more harm. Yale University's *Yale Environment 360* climate change program stated the situation well in 2019:

> *Human intervention with the climate system has long been viewed as an ill-advised and risky step to slow global warming. But with carbon emissions soaring, initiatives to study and develop geoengineering technologies are gaining traction as a potential last resort.*

Geoengineering falls into two very different broad categories. One is focused on reducing the excessive carbon dioxide "greenhouse gas" already in the atmosphere, described as carbon capture and sequestration (CCS) or carbon dioxide removal (CDR). The second is aimed at reducing the amount of heat received from the sun, referred to as solar radiation management (SRM). Examples of techniques in each that are under investigation include:

- Reducing CO$_2$ level in the atmosphere
  - Massive tree planting
  - Direct air capture and removal of CO$_2$
  - Algae permaculture, from small phytoplankton to giant kelp, to sequester organic carbon, reducing carbon dioxide in the atmosphere
  - Synthetic limestone as a building material, for the same effect
- Reducing solar energy received
  - Marine cloud brightening, to reflect more sunlight
  - Injection of sulfur dioxide into the atmosphere to reflect sunlight
  - Mirrors in space to deflect some of the incoming solar energy

These **geoengineering concepts** are described in online Deeper Dive Note #10 (at www.movingtohigherground.com). An excellent white paper about the different geoengineering techniques, including an assessment of their scalability and finance, has been prepared by the independent Foundation for Climate Restoration. Led by Peter Fiekowsky, the organization has the very ambitious goal not just to stop global warming but actually to restore temperature and carbon dioxide levels to the conditions that prevailed throughout most of human civilization. (The report is available at https://foundationforclimaterestoration.org/Resources/.)

As this is a very dynamic field, anyone interested is encouraged to do an up-to-date search for the latest information. None of the techniques seem to be without risks or capable of fully addressing the need to stop the warming. It will likely take some combination of those approaches, with advances in technology. Of the seven efforts bulleted above, my personal assessment is that the cultivation of algae may hold the greatest potential in terms of scalability, efficiency, and impact, with the least risk.

Sir David King (who wrote the afterword for this book) is now a geoengineering advocate. As the former British government chief scientist told a 2019 conference, "Time is no longer on our side. What we do over the next ten years will determine the future of humanity for the next 10,000 years." King helped secure the Paris Climate Agreement in 2015, but no longer believes cutting planet-warming emissions is enough to stave off disaster. He is in the process of establishing a Centre for Climate Repair at Cambridge University. It would be the world's first major research center dedicated to a task that, he says, "is going to be necessary."

PLANTING TREES

Of all the proposals aimed at reducing greenhouse gases, tree planting is the most benign and appealing at first consideration, but the scale needed to make a meaningful reduction of CO_2 is epic. A study published in July 2019 came up with the eye-opening figure that planting the optimal species, in the right soils around the world, could capture 205 gigatons of CO_2 in the next hundred years.[13] That is an amazing two-thirds of all the CO_2 that humans have generated since the industrial revolution and a welcome positive possibility. To achieve that, however, they calculated it would require planting 500 *billion* trees. While the concept is fascinating and worthy of discussion, it needs to be noted that after trees are planted, it would take many years, even decades, for them to be mature enough to have significant effect on converting atmospheric CO_2 into plant biomass. Also, as a reality check, in 2019 the government in Brazil took steps to reverse twenty years of rainforest conservation, encouraging logging and development in their iconic rainforest. Recent extreme fires from California to Australia attributed to the record high temperature also are diminishing the natural carbon sequestration of forests.

Regardless, the tree planting could make a difference and should be pursued. But even that is not a complete solution, underscoring that we need multiple efforts. To keep our perspective, nothing can keep the ocean at the present level given the extraordinary excess heat already stored in the oceans. As we are planting trees and taking other great measures to try to reduce the CO$_2$ and slow the warming, we still need to begin our move to higher ground.

REALITY CHECK: HAS EXPONENTIAL GROWTH STARTED?

In the last chapter we looked at the astonishing effects of "doubling," a form of exponential growth. The latest information from Greenland reinforces the seriousness of the situation regarding sea level rise. On December 10, 2019, a feature in the Washington Post was headlined:

> *Greenland's Ice Losses Have Septupled and Are Now in*
> *Line with the Highest Sea-Level Scenario*

The latest tabulations of ice mass loss show a melt rate seven times faster than three decades earlier. That is just shy of doubling every decade, which would be eightfold. We have no idea if there will be another doubling over the next decade, but it should certainly strengthen our concern regardless of the precise number. The underlying paper in the esteemed journal *Nature* described very high confidence in the analysis, because it looked at twenty-six separate studies. Annual ice mass loss went from 33 billion tons a year in the 1990s to 254 billion tons annually over the thirty-year period. That's essentially doubling each decade. And it's not just Greenland. Antarctica is seeing similar rates of increase. It's a sobering place to end the science part of this book.

WRAPPING UP THE SCIENCE

The warming world, the melting polar ice, and the rising sea pushing us to higher ground are a clear and present danger. Unfortunately their incremental nature permits us to procrastinate in terms of suitable solutions that have sufficient scale. Now that we understand the unstoppable aspects of rising sea level, it's appropriate to consider our full range of options. As we move ahead looking at the engineering aspect and others, this rhyming list of possible actions may be helpful as a top-level structure.

Framework of possible actions to address rising sea level:

- Educate—sharing and teaching in all forms is arguably the easiest action, and perhaps most important, for wider understanding of the new reality
- Advocate, Legislate, and Regulate—push for change and enlightenment of policies
- Elevate—raise buildings and infrastructure to reduce flooding
- Isolate—install means such as levees and sea walls, but only if the geology is appropriate
- Relocate—move to higher ground
- Innovate—embrace the new reality and "think outside the box," e.g., floating or movable structures

"Is this what they mean by adaptation?"

PART TWO

CONNECTING THE DOTS; PLANNING FOR A NEW COASTLINE

5

ENGINEERING WITH A MARGIN OF SAFETY

If a person does not consider the long term,
he will have short-term worries.
—Confucius

Engineers are naturally cautious, but that can manifest in two different ways. It can be a tendency to do things following long-standing procedures and practices. Or it can be a cautious concern to get ahead of a problem when fundamentals are changing. Rising sea level poses just such a stark choice, with the clarity of the fact that it will be many feet higher than during all of human civilization.

My use of the term *engineers* here is in a broad sense. It goes beyond the specific professional designation, and even beyond the related fields of architects and planners. Reengineering our world to accommodate much higher sea level will require the full participation of the law, finance, the military, and all the social sciences. Part of the problem is that scientists and engineers are not communicating well about rising sea level because of two rather different perspectives and cultures. I hope that my broad, simplistic characterization of scientists and engineers here is understood as a means to make a point to the public.

As described in Chapter Three, scientists—more specifically gla-ciologists—are working very hard to understand the forces that will transform the mile-tall ice sheets on Antarctica and Greenland into a higher sea level. At the risk of oversimplifying, scientists' collective focus is to understand and quantify all the factors that will affect the glaciers. Using the work of many international teams on many differ-ent specialized aspects of the issue, they tally the total of the height that sea level could rise based upon the known processes. For consis-tency and comparability, the year 2100 is commonly used as a point of reference for what could happen. Hunches, concerns about tipping points, and potentially sudden "domino effects" of glacial collapse generally do not make it into the scientific projections for SLR until they are quantified with a high degree of confidence and agreed to by a jury of peers. Understandably, scientists do not want to overstate the possibilities or be found wrong. Where there is uncertainty, scien-tific protocol is generally to understate until something can be proven independently and quantified. Somewhat in contrast to that scientific protocol, those who are engineering solutions want to know the real-istic possible worst case, so they can design and construct safe and durable buildings, infrastructure, and equipment. They too work with strict protocols and look for precise numbers, but with a somewhat different perspective.

Where the scientist's training is to say what *will* happen with a given set of inputs and conditions, the engineer wants to know what *could* happen. Engineers typically take a *better safe than sorry* approach. Even use of the defining year 2100 has a totally different effect on the two viewpoints. Scientists need a defining time frame for any finding to have relevance. Saying something could occur sometime in the next few centuries or millennia would be unlikely in science circles, as it is simply too vague. For example, if scientific consensus was that those six glaciers on West Antarctica were likely to add to sea level in the next several centuries, that would not be put in the projection to hap-pen by the year 2100, the standard benchmark for climate studies. In contrast, if an engineer is designing something exposed to sea level, they likely will want to allow for virtually anything that could reason-ably happen within the *design life*. Whether it will precisely happen by midnight on December 31, 2099, would be largely irrelevant.

That should help to understand the cultural chasm between a scientist whose aim is to predict how high the sea will be by a given year, with a known degree of confidence, compared to the challenge of engineering something for the coastline in an era of rising sea level, without knowing the precise rate of rise.

> For those interested, bridging the gap among scientists, engineers, and other professions to have a shared, cross-disciplinary understanding of rising sea level is the focus of my work with the nonprofit Rising Seas Institute, www.risingseasinstitute.org.

Rising sea level will be different from any other global transformation we have experienced to date. For one thing, it is happening slowly, which allows us to perpetuate the illusion that it's not urgent. Second, it is so disruptive that we cannot even wrap our minds around it. And third, it does not trigger either of our main two motivators: fear and gratification.

If there is one "silver lining" to the cloud of needing to move to higher ground, it's that we have decades to do it. Unfortunately, we are delaying and dithering. As will be seen, with very few exceptions we are taking "baby steps" when we should be taking giant strides. We need to think big to prepare for the revolutionary concept of radically higher sea level, which will also raise the height of short-duration flood events. Even that concept of short-term flooding having a different character than long-term higher sea level is a paradigm shift. We need to find new ways to talk about the topic better.

THE FIVE FLOOD FACTORS + EROSION

Our ability to solve a problem largely depends on how we *frame* or conceptualize it. In most coastal areas, rising sea level is seen as flooding. In areas with high bluffs, the problem may be perceived as increased erosion as the sea undermines the cliffs. Even where flooding is "the problem," that is too broad a characterization. It is much more effective to consider the diverse *causes* of flooding, as the differentiation leads to better solutions. I like to categorize five flood factors:

1. Storms and storm surge stemming from hurricanes, typhoons, and other extreme weather events
2. Heavy rainfall, with monsoons as an extreme form
3. Downhill runoff, as well as downstream flooding
4. Extreme high tides, the so-called king tides
5. Sea level rise, including from both higher global sea level and land subsidence

I find these five varieties provide the right framing to consider how to protect ourselves from flooding. The five flood factors can all combine for even worse flooding, but each has different characteristics of timing, magnitude, and permanence.

Note that tsunamis, the extremely high and sudden flood events, are omitted from my five flood factors, as they are caused by neither weather nor melting ice. Tsunamis are caused by seismic events (earthquakes), often deep underwater, which can suddenly generate a wall of water more than 15 meters (roughly 50 feet) in height. As such, they are an entirely different risk than weather events, tides, and rising sea level.

It is worth going into some detail to appreciate the different qualities of each flood type, and their potential to occur in combination, for even greater flooding.

Storms in coastal areas evoke familiar images of the destructive and famous hurricanes, also known as cyclones or typhoons in different regions. These systems of wind, waves, and low pressure deliver vast volumes of saltwater to the shoreline, which becomes even higher in places where the ocean is pushed up into the confined spaces of waterways, rivers, and bays, causing water levels to pile up higher. These waters recede rather quickly, usually within hours, often revealing that wave action has eroded or "re-sculpted" the shoreline. In the coming decades as the oceans continue to warm, the added heat energy will be the driving force behind more severe and more frequent storm events.

Rainfall and *runoff*, my next two items, are related but separate. Both now regularly break records worldwide and are as likely to occur far inland as on the coast. Heavy rain can be many inches a day. As gravity and topography guide the rainfall downhill, water depth can accumulate, turning inches of rain into flooding several feet deep. One way to distinguish rain from runoff is direct and indirect accumulation. Rain's effect is where it lands; runoff can greatly accumulate any distance downhill or downstream, even far from the original location of the rain. As flood factors they are very different.

Record rainfall is becoming commonplace and relates to the warming ocean. Like any body of water, warmer oceans evaporate more. The atmosphere has a finite limit for the amount of moisture it can hold, so excess moisture in the air must come down as rain or snow, depending on the local temperature. Unfortunately, when most engineers, architects, and planners learned their professions, it was assumed that the historical averages and extremes of past rainfall and runoff would reasonably cover the future. Though the weather is always understood to vary from year to year, the frequency and range of extremes from past centuries were thought to be sufficient guides to the future. The estimate of a hundred-year flood was based on the statistical one percent probability of a given amount of rain. Now we are becoming familiar with the new reality that hundred-year deluge rains can happen every year and, in some cases, multiple times in the same year.

All-time records for rainfall and flooding are now broken so often that it's hard to keep track. Previous peak water levels, "high-water marks," have lost much of their value as guidance for safe design. The question we should be asking is: What should we expect in the future? From my vantage point at the start of the 2020s, headlines of vast disaster floods from the American Midwest to India and China are becoming almost annual events. In addition to fatalities and property destruction, there is a huge impact on agriculture, which disrupts food production. The foundation of agricultural productivity is having a reasonable expectation of seasonal patterns—particularly when to expect the rains and their intensity.

Hurricane Harvey, which hit the Houston area in 2017, likely lays claim to the greatest US record rainfall, with some 40 inches landing in just four days, and a total of 60 inches (152 cm) during the ten-day period of the weather system. Harvey is tied with Hurricane Katrina as the costliest tropical cyclone on record at $125 billion in damage. It killed over a hundred people. To illustrate the importance of considering rainfall and runoff separately, the 5 feet of direct rain turned into flooding of double that depth, due to changing terrain and underpasses in the network of roads. Drainage design and engineering had simply not anticipated that volume of water. To make matters worse, lax building codes had allowed developers to turn far too much natural drainage into streets and parking lots, greatly increasing runoff. The catastrophic flooding associated with Harvey made clear the risk of shortsighted development policies that had given developers relatively free rein to transform wetlands and natural drainage into hard surfaces. It is now a textbook case to show the long-term risk of poor zoning and planning.

On August 8, 2017, I was the keynote speaker for an all-day workshop called "Resilient Texas: Planning for Sea Level Rise." The program, sponsored by several state and federal agencies and academic institutions, was appropriately on a barrier island, in a very small town, Port Aransas, not far from Corpus Christi.[14] While complimenting the Texans for their forward thinking, I told the audience the challenge of rising sea level would grow greatly as early as midcentury. That awareness should be incorporated into their land use planning and building codes ASAP, as they considered more imminent forms of flooding, "such as a hurricane, which could be here in a matter of weeks." With uncanny timing, Hurricane Harvey spun up quite suddenly two weeks later, heading for Texas. Back home in Florida, I listened to news of the disaster, worried for those affected, particularly some of my new friends in the disaster zone. When I heard that the extreme hurricane made landfall in a little town, Port Aransas, I was in disbelief. What a bizarre coincidence that it was the exact location where I spoke about just such a scenario.

Runoff from heavy rainfall can be a problem even in low-elevation places like Houston or Miami, but the effects are greatly exaggerated in areas with steep terrain. From the ravines of New Hampshire, to Asia's Himalayas and Hindu Kush, to the mountainous terrain of Hong Kong, rainfall can multiply by more than an order of magnitude (factor of ten) as it gathers volume on its downhill path. Where there is a stream or river, it becomes a natural vehicle to turn the downhill runoff into an even bigger problem, possibly breaching riverbanks and levees hundreds of miles downstream from the area of rainfall.

Extreme tides, my fourth item of the five, is different from the first three in that it's not associated with weather at all. Tide heights are predictable by location at specific times. Tides are a function of the planetary positions and the gravitational pull on the ocean bulge as it moves daily around the globe, affected by the shape of the ocean basins and continents. In the coming decades, the tides may change slightly, but it can be assumed they will be relatively stable following the nineteen-year pattern referenced in the sidebar on the Metonic cycle in Chapter Three. For now, what is getting the attention of communities all over the world is how the extent of high tide flooding is getting worse, even when the weather is good. These increasing floods are often referred to as "blue sky flooding" or "sunny day flooding." Many residents don't even notice that the pattern of flooding follows the lunar cycles. What is being noticed from Boston to Bangladesh and from Vancouver to Venice is that, year by year, the flooding expands in area and duration as the ocean's volume increases. In other words, the cause of higher tides is rising sea level, the fifth item and our primary focus.

Mega-tides are different from "king tides." Mega-tides are daily gigantic tides in particular places. Besides being an interesting oddity, they might help us to think differently about designing to accommodate varying water levels, which will be one of our challenges in the next few decades as we try to build functional and attractive communities. The Bay of Fundy, between Nova Scotia and New Brunswick, Canada, is rather well known for such tides. The daily rise and fall there is some 43 feet (13 meters) with an official record of 54.6 feet (16.6 meters). Its height record is tied with an area a little farther north, Ungava Bay. Both have tides that are a good 5 feet higher than other mega-tides in places such as Mont Saint-Michel, France; Bristol Channel, England; Cook Inlet, Alaska; Sea of Okhotsk, Russia; Patagonia; Panama; north-western Australia; and New Zealand.[15] Mega-tides mostly occur because of the particular shape and contours of that ocean basin, pushing more water into a reduced space.

Compared to the sudden onset and large potential quantities of water involved in the first four flood factors, *rising sea level* is easy to miss. Amounting to mere fractions of an inch (millimeters) a year, sea level rise is the stealth factor. Understandably, we focus on the four more volatile flood factors. It's easy to ignore the melting ice caps adding to the ocean quite imperceptibly. "Boiling the frog" is the popular metaphor these days to describe slow change that can be deadly if it goes unnoticed. Also, Aesop's fabled tortoise winning the race against the hare holds an important lesson; slow and steady can accumulate to overcome large movements.

EROSION IS A BIG PROBLEM, BUT IT'S NOT FLOODING

The "five flood factors" can occur in different combinations in coastal regions. Beach erosion is a growing concern all over the world and is often wrongly attributed to rising sea level. Before looking at solutions to sea level rise and flooding, we need to understand why erosion is something different. Under *storms,* above, we noted that huge coastal storm waves can rapidly claw into coastlines, a sudden and severe form of erosion. That's quite distinct from routine erosion, the natural effect of sand moving along the shoreline on a daily basis. You can observe that at a sandy shore, even on calm days, with tiny ripples moving the sand particles laterally. Where there is a break on the coast, such as a waterway inlet, jetty, or groin, the normal coastal sediment transport is interrupted, often causing massive *accretion* (buildup) on one side, and *depletion* (erosion) on the other. The fact is that coastlines migrate naturally, sometimes inland and sometimes seaward. We ignored the wisdom of our ancestors, who recognized the threat of coastal erosion and storm, generally erecting buildings a safe distance from the water's edge. In the last half-century, we've built closer and closer to the sea, ignoring the obvious risk. Condominium developers, real estate sellers, and "we" as buyers all were lured by the appeal of being as close as possible to the water. Unfortunately, that shortsightedness has led to an enormous problem.

An entire industry of dredging and "beach restoration" now exists and thrives; it's also called beach re-nourishment or replenishment. A half century of building communities next to the beach has resulted in a huge effort to effectively do the reverse: to build beaches next to communities. In some places with high coastal erosion, the restoration is being done very frequently. It reminds me of poor Sisyphus, from Greek mythology, repeatedly pushing the boulder uphill, only to have it roll back down. As erosion accelerates in thousands of places, it has become a battle to prevent properties from falling into the sea.

Hundreds of thousands of properties worldwide now regularly lose their beaches. According to coastal expert Dr. Rob Young, nine billion dollars have been spent on beach restoration over the last century along US coastlines, but the pace of erosion is only accelerating—and rapidly. In Florida, more than one hundred million dollars have been spent in just a few years, some on restoring the same areas only a year or two later. It is a global problem, but also big business for the dredging companies. In beachfront communities from California to the Carolinas and Thailand to the Netherlands, huge sums are now spent keeping beaches where we want them. Decent-quality sand is already in short supply, making beach restoration increasingly expensive. Local politicians, homeowners, real estate developers, and the broad swath of those involved with beach-based tourism are consumed with finding funding for beach building in coastal communities worldwide.

To be clear, coastal erosion is affected by rising sea level but is fundamentally different. That's why I frame the category as "five flood factors + erosion." As already described, we cannot know with any precision how high sea level will be, even just thirty years from now. Farther out, the range of possibilities for sea level gets much wider.

Adapting to such ambiguity in advance presents a very real challenge. Nonetheless, there are some ways to look at this positively, which is the intent of the next few chapters.

- The clarity that sea level *will* rise and is unstoppable should enable and inspire bold planning.
- Even if we cannot know the precise rate of future rise, planning for higher is safer. Extra height above the water level is sometimes referred to as "freeboard," borrowing a nautical term.
- Those who can see the trend early may profit, or at least can reduce their risk and losses.

ADAPTATION IS COMPLEX

Motivations to boldly plan for the future can encompass humanitarian and social concern, pure problem-solving, avoiding loss, or creating value. Raising streets, buildings, ports, and infrastructure is difficult and costly work. One thing impacts another, and the effects can be very disruptive. For example, if one street is raised, the water flows elsewhere, typically towards houses, stores, and other parts of the community, perhaps making things worse and causing liability issues. Drainage criteria have to be looked at on a large enough scale to consider the new levels of rain and runoff, urban density, permeable areas, and rising sea level. There are already places where rain runoff that had been allowed to run to the sea or to a waterway by gravity now has to be pumped. As sea level moves ever higher, even many of those pumping arrangements will not function. Coastal infrastructure that is being designed, approved, or constructed today should anticipate the new realities in order to have a longer useful life, thereby having a better ROI, return on investment.

Deploying sandbags, using pumps to keep the water out, and evacuation may be the only option if a flood is threatened "today or tomorrow." With each flood event, or even as the waters rise higher, threatening to flood, our attention turns to what can be done to reduce the risk of future flooding, often referred to as *flood hazard mitigation*. Raising streets and homes, one by one, a foot at a time, can be a short-term response. However, even elevating a house or commercial building will do little good without a plan to keep the community functioning. The desirability and value of a property rests on the availability of utilities, emergency services, grocery shopping, retail, jobs, transportation, and amenities. In the face of accelerating SLR, I recommend doing a thirty- or hundred-year master plan at the community level. It should include zoning, building codes, and infrastructure that anticipates some level of future SLR. Without a cohesive plan that addresses the issue at the community level, even the best efforts will end up underwater.

It is time to start engineering for a sea level that will be dramatically higher. For potentially vulnerable assets, in the face of unstoppable rising sea level and flooding, it should be obvious that there are four primary approaches to consider as solutions:

1. Keep the water out with some form of high walls. (Issues: How large an area? Does it need to be aesthetically appealing? Is the geological structure sufficiently impermeable?)

2. Elevate structures above the water, typically on pilings or by building up the land height. (Issues: What about community services, utilities, and sense of community? Scale?)

3. Use some form of floats to adjust with the rising tide. (Issues: Challenge of marine corrosion, exposure to severe storms, and safe utility connections.)
4. Move to higher ground, the most scalable, economical approach for most vulnerable people and communities in the long term. (Issues: Reluctance to abandon place/community, both from economic and emotional viewpoints.)

Costs and obstacles will vary greatly depending on location and who is asking the question. Suffice it to say that, in practical terms, as a society our adaptation will be a combination of all four. We need to maintain our presence and access to the coast for the myriad of reasons that have caused people to build near water for all of human history.

Theoretically, there is a fifth option: to live below sea level, underwater. However, it's not realistic to extend our domain underwater for even a fraction of the millions of people threatened by the rising ocean. There will be underwater facilities for research, mining, and industrial purposes, but due to the pressure, ocean forces, and corrosive environment, as well as costs, living below sea level is not on par with the four options listed.

LESSONS FROM THE DUTCH

The Dutch are often looked to for leadership in coastal flooding, and for good reason. An estimated twenty-five percent of their country is actually below sea level. One of their favorite sayings recognizes that a large part of their country is land that was naturally underwater:

> God created the Earth, but the Dutch created the Netherlands.

For a thousand years, the Dutch have been expanding their small nation with dikes, the earthen walls that can keep water out. Dikes require enough clay in the levee material to make it impermeable and for the deeper geologic structure not to be permeable either. With those criteria, dikes can raise the shoreline both on the ocean and on rivers. Once the *dikes or levees* are built, water can be pumped out, leaving what the Dutch call a *polder,* a dry area below sea level that can be used for farming or communities. Originally windmills pumped the water, eventually giving way to huge electric pumps. Positioned at the delta outflow of three major rivers—the Rhine, Scheldt, and Meuse—Dutch soil is rich from the river silt. That soil has been the basis for agriculture and flowers, mainstays of their economy, with worldwide exports.

A single tragedy defines modern Dutch history, the sudden North Sea Storm of February 1, 1953. During the night it broke through the dikes, killing 1,835 residents. That was the impetus to completely reevaluate and reengineer the Dutch coastal defenses, which became the *Delta Works program.* The semicircular twin gates at Rotterdam, the Maeslantkering, became an iconic engineering marvel to prevent future storm surge from entering their main port and the river system. It was designed in the 1980s for what was thought to be a worst-case scenario. There were three main design criteria. The engineered solution had to cope with a once-in-ten-thousand-year storm, plus the maximum projected downstream flooding from the three rivers that joined there, plus 30 centimeters (about a foot) of higher sea level. Four decades ago, that was thought to be the worst-case scenario for SLR this century. Back then, engineers were not looking at the possibility of sea level becoming 10 feet higher. Dutch engineers now acknowledge that with the acceleration of polar melting causing faster sea level rise, the gates should have been designed 2 or 3 meters higher, to have an adequate factor of safety.

The point is that good planning needs to use long-term criteria so that structures are designed and constructed to properly function when they need to. Because far higher sea level is not within our human experience and frankly is something we prefer not to contemplate, our tendency has been not to design for it.

Sand Motors and "Room for the River." While we're on the subject of Dutch engineering and innovation, two other concepts should be explained, as they are getting public attention but are somewhat misunderstood. The first "sand motor" (or sand engine) was created in 2011 on the coast of South Holland. It consisted of piling over twenty million cubic yards of sand at the right location so that the sand would be distributed by ocean currents to areas of high beach erosion farther down the coast. While it is an efficient way to do beach restoration on a broad scale, it can only work in certain locations. Also, it is still very expensive and should not be seen as a widespread solution to erosion.

"Room for the river" has been described in the popular press as another "solution" to flooding. It creates a "relief valve" to allow high water levels to leave a river either by floodgates or a designated lower height in a section of levee. It requires the purchase of property or easements to allow large tracts of land to be flooded selectively in order to protect communities downstream on inland floodplains. The Dutch have optioned some large agricultural areas for this purpose. In addition to the Netherlands, this important tool is used on the Mississippi River, in the UK, and in Egypt, China, and India. As river flooding continues to increase, due to deluge rain associated with a warming ocean, it is important to design such sacrificial flood zones. Just to be clear, however, **room for the river** does nothing to prevent coastal cities from flooding as the ocean rises.

We also need to recognize that the Dutch have a great cultural distinction that cannot be transplanted to the United States or most other nations. Public policy in each nation must be shaped and supported by the people and their elected representatives. Even with all the flood disasters in the United States in the last decade, we have none of the deep Dutch awareness that comes from their thousand years of dependency on levees and from the horrific 1953 tragedy. From direct experience, they know they can suddenly be flooded, subject only to the security of their dikes. Different places and cultures prioritize their top-level threats very differently. The Netherlands and Singapore have made rising sea level their highest-level threat this century, recognizing the profound implications for buildings and infrastructure. Presently, most other nations lag far behind.

PLANNING: BEGIN WITH THE END IN MIND

There are two reasons we should start thinking bigger, sooner. The first is practical. Planning farther ahead and raising streets and structures 3 feet one time is generally less disruptive and less expensive than raising them 1 foot three times. Applying that thinking to roads, utilities, and infrastructure, it's far more efficient to make the changes once, rather than to keep making smaller modifications.

The second reason to plan for much higher SLR sooner is that it sends a stronger signal to the wider community, as well as to ourselves, about what lies ahead. If we just keep raising buildings, roads, and infrastructure inches (centimeters) at a time, we continue the illusory perception or culture of denial. Such "incrementalism" misses the opportunity for large-scale design change. It's like the difference between continually modifying a building versus demolition and designing from the ground up. If we are ever going to be ready for a sea level that is 7 feet higher (2 meters), we have to recognize the reality of a new baseline and think on a different scale altogether.

Decades ago, The *7 Habits of Highly Effective People* was a best-selling book by Stephen Covey. Habit number two was to begin with the end in mind. His analysis showed that high achievers often start with their eventual goal, which might be considered "point C." With that identified, the intermediate steps, "points A and B," become much clearer and easier to achieve. In other words, having a very clear vision of the desired far-sighted situation makes it much easier to define and adhere to the steps to get there.

Whether one is designing a building or a transportation system, there is a huge advantage to having the master plan done first, before designing, engineering, and constructing individual elements. Master plans for constructing anything large require detailed analyses that include a wide range of criteria, including bedrock, soil porosity, drainage, power availability, water, wastewater, parking, traffic flow, economics, marketing, and regional demographics. That is all developed and engineered on an iterative basis, ultimately with a model, visualization, or rendering to get stakeholder buy-in.

A cohesive and comprehensive approach is usually the best way to get the job done well. The owner of a house, a restaurant, or commercial office building that floods regularly might push for the most immediate solution to that property's flooding problem, eager to avoid inconvenience, disruption to their access, or negative impact on the aesthetics of their property. Those owners are likely unenthusiastic about a more dramatic overhaul. If the entire street needs to be elevated to install a new drainage system, it will likely gain support only if there is a comprehensive vision of how it will lead to improved values in the future. Galveston, Texas, is a good historic example of bold planning to get ahead of a problem somewhat similar to the Seattle story at the start of this book. In 1900, a hurricane swept over Galveston Island, killing more than six thousand people. Realizing it could well happen again, the city built a now-famous seawall 17 feet tall, changed the building code, and elevated a large sector of buildings by the same 17 feet. That made the community far less vulnerable, giving them a century of security and prosperity. It would not have been possible for individual property owners to tackle that; it required bold vision, a long-term plan, and leadership to make it happen.

Compare those bold, visionary changes to the modest solutions now being planned, discussed, and implemented. With some examples to be shared shortly, typically the focus is to raise a road, seawall, or parking lot as little as possible. I am not trying to minimize or ignore the obvious difficulties to do good planning and adaptation. On a positive note, we are starting to see articles in the business and popular press about the huge challenge of flooding and sea level rise. A good example was a 2019 feature in the New York Times: "With More Storms and Rising Seas, Which U.S. Cities Should Be Saved First?"[16] As the author, Christopher Flavelle, noted, the unmistakable reality of worse and worse flooding pushes up hard against economic and political realities.

WATER SQUARES, WATER STORAGE, AND SPONGE CITIES

Although short-term flooding is of a different nature than sustained higher sea level, it is relevant since, for one thing, SLR will make it worse. Second, some of the techniques being used to plan for flooding may suggest lessons for the long-term challenge ahead. Some approaches are very creative. In the Netherlands, public parks such as Rotterdam's Benthemplein have been developed as attractive, useful "water squares," places that can function during good weather as community gathering places, with recreational facilities and outdoor artwork. Thanks to their clever design, they become short-term water storage during deluge rain. A large multi-level underground parking facility is also designed for dual functionality for temporary water storage. These solutions successfully divert extra water from flooding streets and homes. No doubt the fields of landscape architecture and drainage engineering will become even more creative in the years ahead.

As reported by the World Economic Forum, China is showing leadership with the *sponge cities* initiative it launched in 2014. Modeled on engineering concepts used in India and Vietnam, sponge cities aim to have eighty percent of all urban land able to absorb or reuse seventy percent of stormwater. More than thirty cities are currently part of the initiative, including Shanghai—one of the most flood-prone cities in the world. China expects that at least another 600 cities will join with their own sponge city designs in the coming decade.[17] To be clear, identifying that Shanghai is on a list of sponge cities in no way means they are on track to a viable solution to address rising sea level more generally. By all reports the world's busiest port, Shanghai has a terrible unresolved challenge as they deal with present-day and future flooding.

NATURAL SOLUTIONS

It's important to consider the appealing natural solutions that are available. The previous chapter mentioned a study that proposed planting 500 billion trees to significantly slow the rate of warming. Such natural solutions can definitely help, but must be considered in light of their practicality and scalability.

In the coastal zone, many people are advocating the use of wetlands, marshes, mangroves, and oyster beds to absorb storm wave energy and provide a more adaptive coastal junction than concrete seawalls. Those "soft structures," often described as *horizontal levees* or organic shorelines, can also serve as buffer zones moderating the usual present-day push for ever more coastal concrete. All those approaches make sense and are terrific. I endorse them wholeheartedly. Yet the sober truth is that those efforts will not be enough on their own. They need to be done in concert with the complex work of designing buildings, communities, and infrastructure that can function in a world where sea level is *at least* several meters (5 to 10 feet) above present.

OUR TIME HORIZON IS KEY

For us to begin adapting in preparation for markedly higher sea level, our "time horizon" is key. Do we want things to be safe and operational for a decade, for a century, or longer? Given that flooding from rising seas and abnormal rainfall is beyond the experience of all human civilization, and with the uncertainty of the melt rate in Greenland and Antarctica, how do we design? It's a challenge unlike most design and engineering situations. We need to consider each location and project, what will likely happen, and what could happen. In a dynamic unstable situation, one's choice of time horizon will greatly affect the design solution. There is no "right answer."

For example, often I hear implied and even explicit reference to a time frame of five years when it comes to planning. That has come from elected officials, apparently thinking of their term in office, but also from professional planning staff who may not have been given the resources or the mandate to plan for decades. Other people with five-year contexts include real estate developers who may think they can buy a property, develop it into something of greatly increased value, and sell it within five years or so, and individuals buying a home or condo, concerned about future flooding but assuming they will have time to sell. Five-year planning may consider short-term flooding but would not include any noticeable SLR.

While five years might seem reasonable for a particular purpose, it misses three problems: First, plans change; it's common to be in a house for twenty years when the intent was to be there for just a few. Second, given the growing awareness of flooding from all the factors, property values are already being discounted in anticipation, a trend we can expect to increase. Third, those short-term views simply prevent us from looking at the long-term trend, the sweeping arc of change and the growing chance of something happening abruptly within a few decades. Of course, there are situations where it is good to look down to see where you are taking the next step. However, that should not preclude looking up to see what looms ahead in our future path.

To see the importance of the right time horizon, let's consider a hypothetical engineering evaluation of a bridge, from three different perspectives. First, an insurance company asks an engineer if the bridge that they want to insure is structurally safe during the next twelve months, the length of their insurance policy. Second, a company financing or approving a bond issuance for the same bridge would want to have it evaluated for thirty or fifty years, which may result in a very different assessment of its safety, durability, and cost to maintain. Third, let's say it's a toll bridge, to be completed through one of the popular public–private partnerships being used to finance infrastructure, often called a "P3." Their business model is for private investors to spend tens of millions to build the bridge and then share in the tolls collected, perhaps for a hundred years, doing good maintenance like repainting, sandblasting, and inspections.

The engineering analysis of that same bridge will be quite different for the insurer, the financier, and the owners. Looking out over one hundred years, the concerns for points of failure and risk will include many issues that are irrelevant for the one-year insurance policy or traditional thirty-year financing. Same bridge, same engineer, but three completely different engineering evaluations. The same changing perspective will apply to most coastal assets as we consider the future with seas rising to new heights.

In the case of bridges, it should be pointed out that there is one very special design consideration due to rising sea level. Even setting aside the question of how higher water level might put the access road or the bridge's structural support underwater, there is the critical issue of bridge clearance height for cargo ships, cruise ships, and sailboat masts to pass safely underneath. Various tides, waves, and stormy weather must be considered in the thirty- to hundred-year evaluations. How much sea level rise should be assumed? Given that some ships pass with the barest of clearance under bridges, it is particularly important for engineers to use a farsighted planning horizon for bridges, which can easily be in use for centuries. We can no longer design based on the historical hundred-year or five-hundred-year water levels. Apply this thinking to any number of other structures: a tunnel entrance, a chemical plant or oil refinery, or the cooling system for a power plant, perhaps even a nuclear one. Inches matter.

ENGINEERING WITH A MARGIN OF SAFETY

Civil and mechanical engineers need to consider water levels for a wide range of structures, including bridges, buildings, and industrial plants on the coast. In the UK, the most iconic is the Thames Barrier, designed to protect Greater London from flooding, which has been used with increased frequency in recent years. The set of eleven floodgates, operating since 1982, have now exceeded their design criteria due to rising seas and must be put into a hyperextended position for peak events. The question is what criteria to use in designing their replacement, which should anticipate a further century of use to protect the city.

Addressing the larger issue of safe guidelines to design for future sea level rise, the UK's Institution of Mechanical Engineers (IMechE) set up a task force in 2018 and brought me on as their subject matter expert. Over the course of a year, many workshops and meetings were held, spanning various sections of expertise among their hundred-thousand-plus members. Traditionally, most design criteria are based on the premise that the recent past is a basis for the future. Historical ranges for the last 100 years, or even 500 years, are assumed to cover 99.8 percent of occurrences. Recognizing that climate change is changing the variables, engineers are increasingly trying to plan for sea level rise. The problem is that typically they reference the IPCC report described in Chapter Three without realizing its limitations. Recall that its most extreme scenario had a sea level rise of 32 inches (90 cm) by the year 2100. As I pointed out, and as the US Department of Defense study noted, those figures essentially omit the contribution from the melting of Antarctica only because they were not able to quantify it with sufficient precision to meet their own requirements.

With the British engineers I used an analogy for a factor of safety. For example, when someone is designing a structure like a bridge, the steel is not engineered just to *meet* the load requirements. Rather, they start with the *maximum* number of cars and trucks, with an assumed maximum average weight per vehicle. From that they calculate an estimated maximum weight load. They then add an *additional* "factor of safety" to that figure, in some cases more than double the projected load. No bridge or structure would be engineered just to meet the actual stress calculation; there is always a huge safety factor. Yet we are not doing that with rising sea level at all. Planning for SLR almost universally takes the projections, including the Antarctic minimums, and designs to the number. There is no significant extra safety factor.

Thus, I was delighted that in late 2019, IMechE issued an excellent report, "Rising Seas: The Engineering Challenge."[18] It is a breakthrough in terms of how to design in the face of unpredictable, potentially abrupt, and permanently higher sea level. The report strongly makes the case to plan responsibly for SLR that could well be meters higher over the coming century. To my delight, they embraced a simple tool I developed, shown below, that introduces a margin of safety. They refer to it as "Englander's 9-Box Matrix," though perhaps something more official and auspicious will evolve as a title.

"9-Box Matrix" – Planning height guidance for sea level rise, plus a margin of safety			
Risk Sensitivity	30 Years*	50 Years	100 Years
Low	30 cm (1 ft)	60 cm (2 ft)	2 m (7 ft)
Medium	60 cm (2 ft)	1.3 m (4 ft)	4 m (13 ft)
High	1 m (3 ft)	2 m (7 ft)	6 m (20 ft)
* Reference year for projection = 2020, i.e., first column is approximately the year 2050			

Figure 8. The latest simple information sheet with the matrix and a graph are available at the website of the Rising Seas Institute: https://risingseasinstitute.org/englander9boxmatrix/. Note: this may be updated from time to time.

To be clear, these numbers are just meant as guidance for professionals to evaluate and apply and will surely have to be tempered for different situations. Some projects, such as a recreation area, will be able to reference the low-risk sensitivity line. Others, such as a new nuclear power plant, would likely want to consider the high-risk sensitivity line. The IMechE analysis and report should serve as a model for other similar professional societies. Rather than focusing on how high sea level will be, the approach is to provide engineering guidance with a margin of safety over the full range of what is possible.

THE THIRTY-YEAR PERSPECTIVE

Notice that the first column in the 9-Box Matrix is thirty years. I have come to believe that's the ideal time horizon for most considerations of future flooding. It gets us to midcentury. It's a mortgage period or "building cycle." It also gets us into the era when sea level is now projected possibly to be more than a foot higher. All that, combined with Covey's second habit—to start with the end in mind—makes the thirty-year perspective very useful. Typically when issues of flooding and climate change are considered, the reference point is the next decade or two, or the end of the century. I find that less than twenty years is too short to see potential impacts of rising sea level, and the year 2100 is so far in the future as not to seem relevant to most people.

Thirty years is perfect to look at enough time for effect and still to seem within reach. It is useful for the public as well as professionals. Professionals will also want to add the fifty- and hundred-year benchmarks for planning and design.

ADAPTIVE ENGINEERING

Finally, I want to encourage a concept that is starting to be used in the UK, the US, and elsewhere. *Adaptive engineering* describes an approach to allow for future changes. As a simple example, imagine that you were building a seawall but were not sure whether to build it to allow for 1 meter or 2 meters of SLR during its projected hundred-year useful life. If you designed the foundation adequately and anticipated a future *add-on,* if needed, you could start with the lower height and modify it later at far less cost than having to start from scratch with a new foundation. Adaptive engineering will not solve all issues of unpredictable and accelerating SLR. Almost always there will be limits to the adaptation, dependencies on other issues beyond your control, and the larger context of how communities and regions will be forced to change. But it represents an important departure from thinking that things are static.

Having reviewed the problem of rising sea level and the essential need for adaptation, we will now turn the page to *different places and different solutions.*

6

DIFFERENT PLACES, DIFFERENT SOLUTIONS

You never change things by fighting the existing reality. To change something, build a new model that makes the existing model obsolete.
—R. Buckminster Fuller

"Miami has a big problem with sea level rise" is something I hear often, obviously with some validity. Unfortunately, such a characterization has the effect of making SLR "not my problem" for every place other than Miami. Pointing to others in a worse situation is false comfort at best and misleading at worst. Rising sea level will directly and indirectly affect all coastal communities from Providence to Los Angeles and from Vancouver to Sydney—the list is almost endless.

Infrastructure and industry like seaports, coastal airports, rail terminals, refineries, and regional utilities, particularly nuclear power plants, will have larger impact zones. They will even affect homeowners who may be a hundred feet up on a bluff, safely above direct submergence. Those homeowners may not have to abandon or relocate their houses due to frequent flooding or submergence. Nonetheless, they may be impacted as key coastal facilities are under threat of increased flooding as the water rises.

If we are going to tackle SLR, we must frame it accurately as a direct threat to the twenty-four US coastal states and 150 nations with ocean exposure. In today's globalized economy, supply chains and distributed business holdings will spread the impact of rising sea level to most people even if they live far inland. From the private equity firm in Austin to a bank in Zurich, more people will have their assets exposed to sporadic flood events and long-term rising seas than might be commonly realized.

Florida is a prime example of the complexity of vulnerability to SLR and the need to assess with some detail. Contrary to popular impression, Florida is neither flat nor about to "slide beneath the waves." For example, the tourism mecca of Orlando is a solid 80 feet above sea level and not in any direct jeopardy from rising sea level. There are dozens of small communities on Florida's high plateau, almost two hundred feet above sea level, that have virtually no risk from the threat. They dot along an ancient coral reef ridge, running through central Florida. The capital, Tallahassee is a full two hundred feet of elevation. Of course there are a vast number of Florida coastal cities and towns that are at water level and extremely vulnerable to rising seas.

The point is that for a problem to be tackled, it must be accurately defined. As estimated earlier, more than ten thousand coastal communities all over the world are going to have to face the hard truths about adaptation. Though we tend to think first of large cities, the smaller cities, villages, rural areas, and agricultural regions will all be exposed to transformation by the rising sea. Since hardly any two places share precisely the same features, each will have to understand their unique vulnerability and develop their own solutions.

Rising sea level will directly affect every parcel of property that is connected to the oceans with tidal water—that is, water that is even slightly connected to daily ocean tides. That means cities not usually thought of as being on the ocean, such as Washington, DC; Houston; Sacramento; Troy, NY; Montreal; London; and Hamburg, which are all on tidal rivers and waterways.

Major ports present a special opportunity for adaptation, with their vast concrete wharves and intermodal connections with railways and roads. Looking ahead to the second half of this century, they may be able to put into place long-term adaptations to be prepared for rising seas more easily than a lot of facilities. Indeed, they *need* to plan decades into the future, given their vast infrastructure. Fortunately, they are not driven by aesthetics to the same level as residential and retail areas, and thus are able to take a more practical, even industrial approach. The world of shipping goods is hidden for most consumers but essentially affects everyone. Though Miami tops the list of cruise ship ports, it is not even among the top fifty ports for cargo shipments, a list topped by Shanghai, Singapore, and Shenzhen. More than ever, we are all dependent on the uninterrupted global supply chain. Billions of dollars are spent each year to improve those facilities and to maintain them as a competitive logistical chain. The companies involved in port operations and facilities should be among those looking far ahead so that they can "rise with the tide" and continue to benefit from being on the international supply chain. In our global economy, the supply chain has direct effects on family budgets, city vitality, and the share prices of affected companies. As such, the effects of sea level rise will be felt even in cities far from the sea.

Arguably, the century-long strategic view to plan for rising sea level may be more easily embraced by countries that have existed for millennia. Places and cultures with a long view of history may well have an easier time planning for the distant future. One example of this kind of century-level thinking is China's new Belt and Road Initiative, a vast network of roads, rail, and ports designed to bring natural resources and components to China and then relay their output to the world.

If anyone doubts the impact of a port operation's efficiency and reliability on the economics and vitality of a city, they should consider what happened with the introduction of 20- and 40-foot standardized shipping containers a half-century ago. It's hard to remember that goods once moved in boxes on pallets and were loaded with forklifts and cargo nets. With the "containerization" revolution, shipping moved to specialized ports with the right equipment. The wharves of San Francisco and New York City were each substantial cargo ports. With the invention of the efficient steel boxes handled by giant cranes, shipping and supply chains moved from those cities to Oakland and Newark–Elizabeth, respectively, with a huge economic impact on all the cities involved. Similar scenarios played out all over the world.

The point is that planning ahead for SLR will enhance or undermine future economies depending on how communities approach the issue. Those that think strategically and engage in robust long-term planning will almost certainly be rewarded at the expense of those that think short-term and narrowly.

Options for dealing with SLR will necessarily vary from place to place and must be addressed in ways that are rarely considered. Jacksonville, Florida's largest seaport, lies in a vast low-lying delta that is dangerously prone to flooding (as Hurricane Irma demonstrated in 2017). In sharp contrast, Manhattan rises sharply out of the water, is made of solid rock, and can be protected against 10 feet of higher sea level. Much of the surrounding area, however—Long Island is a prime example—is at lower elevations and composed mostly of sand and gravel that will be much harder to protect. Rural locations with low economic value and density will make it harder to justify expensive protection. Compare the cost/benefit of protecting a mile of Boston waterfront against rural Cape Cod, which is essentially a sandbar. In each vulnerable location, solutions will need to be evaluated and prioritized against others.

Boston is highly vulnerable and has done one of the most farsighted studies looking ahead to the era when sea level will be a meter or two higher. There are lessons from its journey to find a solution. One of the earliest radical designs was done in 1988 and called "Boston's Safety Belt." DiMambro and Associates came up with the award-winning concept to connect the barrier island communities into a giant floodwall. It now appears that it was too bold, too early, and was never pursued. Two decades later the problem was studied afresh by a robust team pulled together by ULI (Urban Land Institute). Their 2014 report, "Living with Water," looked at four distinct parts of the historic city to consider how people, land, and buildings might boldly adapt to a new functionality and design with water levels as much as 5 feet higher. One example of their highly innovative adaptations involved converting some streets into canals. In spite of an extremely professional and diverse process and presentation, that report quickly encountered resistance due to the cost, disruption, and difficulty of implementation.

Within three years, however, after lots of discussion by different civic groups and the involvement of a Green Ribbon Commission, the mayor proudly released a graphical and easy-to-read report called "Climate Ready Boston" that included many of the previously proposed concepts: raising specific seawalls, roads, and infrastructure; changes to building codes; and zoning changes. Now there is a robust dedicated website addressing the issue.[19] Boston's example illustrates that adapting to rising sea level is anything but a straightforward process. Even a great report is just the first step. Implementation will not be cheap or easy. In addition to the engineering and expense, the social and political resistance will be daunting. Boston's path underscores that, particularly in democracies, solutions have to take into account not only community-specific geography/topography, but also infrastructure, population and neighborhoods, economic hubs, and culture. Even with that, it's not going to be easy or straightforward. There will almost inevitably be resistance. Solutions to rising sea level require extraordinary leadership, a topic for the final chapters.

Beyond geography and topography, geologic structure can be an important factor. Areas from South Florida to coral-based islands in the Bahamas and the Caribbean, Pacific, and Indian Oceans are largely made of highly porous limestone that defies protection with seawalls, as the water will just move through the rock. (Picture a hard calcium rock with holes and channels like a kitchen sponge.) Water just bubbles up from the ground through the holes in the substrate.

In such porous places, to avoid the dreaded concept of retreat, options will need to focus on raising structures, raising the land itself, or even creating floating platforms. But there are challenges and limits to such approaches. In addition to the land flooding more frequently, in many places the drinkable water will be affected too, as saltwater creeps or intrudes into the freshwater aquifers. In most coastal areas the prevailing perception has been that flooding correlates with closeness to the coast, since the biggest threat traditionally came from storm waves. Whether the water moves in through streams or developed canals, or through the marsh, swamp, or porous rock, the slow infusion from rising sea level is entirely different from storm waves hitting the coast. Over time, water will work its way inland to find and settle into the low spots unless they are surrounded by clay or impermeable rock.

Surprisingly, in many coastal areas, the lowest property is not located right on the coast. In much of the southeastern United States, a bluff exists, just inland of the sand dunes above the beach. Moving inland, the elevation falls off rapidly to either an intracoastal waterway or marshland. This is true for most of the area from Miami north to the Carolinas, up into the mid-Atlantic and a good part of New England. Those low-lying inland areas can make exposure to sea level rise worse. For example, Palm Beach, located on a barrier island, is vulnerable to crashing waves from a hurricane, but sea level rising a foot higher does not represent catastrophe since the famous island is mostly 5 to 10 feet above present sea level. Ten miles west, moving out towards the Everglades, the opposite is true. Storm waves will not reach there for centuries, but even one foot of sea level rise will cause flooding for the many properties bordering the brackish Everglades, which is essentially at sea level. Even many residents are unaware of this paradox.

The point is that localized regions warrant more specific examination. Without question, South Florida is highly vulnerable, but not all areas are equal. An overlooked part of Florida is the historic quaint inland towns with abundant lakes high above sea level. Sebring, Lake Wales, Lakeland, and Ocala are just a few examples of Florida communities a hundred feet or more above sea level. Not only could such places retain their property value, that value may increase substantially as the more popular low-lying coastal areas devalue with increased flooding and people start seeking higher ground without leaving the state.

NEW ORLEANS: JAZZ, MARDI GRAS, AND FLOODING

While most communities are special to their local residents, there are unique cases with national and even international value that arguably warrant extra effort to save or at least to evaluate. In particular there is concern for the historic preservation of hundreds of priceless places. New Orleans, St. Augustine, Annapolis, and Newport (RI) are just a few examples in the US, with hundreds more worldwide. Rising sea level will evoke vastly different priorities for different people. Almost any community will try to hold on in the face of what may at first seem to be bad luck with the weather, until they realize that it's part of a larger pattern that defines a new era.

Efforts to protect and preserve New Orleans demonstrate how far we might go, even in the face of an increasingly futile situation. Emotionally, we simply do not want to let go. New Orleans is a wonderful, unique city, most associated with jazz, Mardi Gras—and flooding. It's built on sinking soils at the mouth of the Mississippi River in the hurricane-prone Gulf of Mexico. After Hurricane Katrina in 2005, nearly 300 billion dollars was committed to restoring the city and fully upgrading the flood defenses. For those unfamiliar with its position, New Orleans is entirely dependent on earthen levees, concrete floodwalls, and pumps to keep the water at bay on two sides, the Mississippi River and Lake Pontchartrain. This 2019 headline in *Scientific American* stopped me on sight:

> *After a $14-Billion Upgrade, New Orleans' Levees Are Sinking: Sea-level rise and ground subsidence will render the flood barriers inadequate in just four years.*

The article revealed that just eleven months after the US Army Corps of Engineers (USACE) completed one of the largest public works projects in world history, the very same agency published an amazing announcement in the official *Federal Register*. The recent repairs and upgrades could fail to provide adequate protection in as little as four years because of rising sea levels and shrinking levees. In remarkably clear and succinct language, the USACE officially announced that it was "doing a full reevaluation of the just-completed project to determine if protecting New Orleans from a 'hundred-year storm' was even technically possible and economically justified."[20] They cited that sea level is rising even faster than they calculated back in 2007 when they did the design and engineering. I found it an amazing state of affairs from several perspectives: (1) how much we had invested to save it, (2) how soon the fourteen-billion-dollar fix may fail, and (3) the USACE publicly stating that a study needed to be done to determine whether New Orleans could be protected at all and whether it was economically justified. While it is tragic, jolting, and poignant, I see it as a sign of progress that professionals are taking SLR seriously and trying to do realistic assessments.

The Mississippi River is the largest river system in the US. The vast drainage basin and river network were generally created about ten thousand years ago, as the last glacial ice sheets melted in North America and sea level reached its present height. The mouth or delta of a large river like the Mississippi typically exhibits natural "delta switching" about every thousand years, as a result of sediment buildup, causing the river to find another path to the sea. But since we like rivers to stay in one location to suit our communities and commerce, we try to tame rivers, keeping them in place. That's the purpose of most of the USACE's work over the last century. It's no surprise that we try to engineer and build infrastructure to confine and redirect the forces of nature. Largely that is what human ingenuity has enabled, allowing our society to flourish. However, as flooding increases and SLR continues to accelerate, there is a limit to what we can do to save particular properties. A rising and accelerating ocean level is a challenge that humanity has never before faced. At some point, we need to start taking the long view and ask where things are headed. How long can we keep the particularly vulnerable coastal cities where they have been for recent centuries? New Orleans is a prominent example that should give us pause and make us realize that we really are in a new era.

INTERNATIONALLY, VENICE AND JAKARTA ARE GOING DOWN TOO

While we may consider places like New Orleans irreplaceable, we can't keep such sinking coastal communities in their current configurations and locations forever. The fact that it's sinking faster than any location in the US may make it a good place to start to confront the mature decisions of intelligent adaptation, including options of retreat or total redesign and rebuild. Internationally, Venice and Jakarta are good comparable examples of cities where high rates of land subsidence make pouring more and more money into temporary fixes increasingly futile. Although both cities have attracted vast sums and creative schemes to keep them afloat, there is no substantive sign of reversing or stalling the long-term trend of SLR that is permanently altering them.

Venice has spent billions of dollars on the MOSE barrier, a promising yet limited solution to rising waters. It's a barrier of 78 units that can be filled with air to rise into position as a temporary wall holding back the Adriatic Sea. It finally passed operational tests in 2020 hopefully, protecting the iconic city when there is a storm or peak high tide. It's not a long-term fix, however. Aside from the limits of its design height, when the gates are in position, they confine sewage in the lagoon. Venice has no wastewater treatment system. The close network of pilings on which the city is built make it virtually impossible to create one now. Presently, the world-famous city known for canals, gondolas, architecture, art, and Murano glass is becoming identified with frequently flooded plazas, terrible stench, and the risk of disease. Saving Venice evokes a similar emotional reaction to that of saving New Orleans. These are places we love. But it is rapidly becoming clear that there are difficult realities that conflict with our desire to keep them alive forever and at any cost.

In the United States, New Orleans is not alone in being super vulnerable due to subsidence and the limits of levees. Areas with particular vulnerability include the region around Norfolk, Virginia—also due to abnormal subsidence—and Sacramento, California, due to aging levees. As depicted below, there are over 29,000 miles of levees in the US with an average age of fifty-five years. That means they were designed and built before the concern about rising sea level and the record rainfall now associated with warming. Already many of the levees are on the verge of failure; more will be in the coming decades.

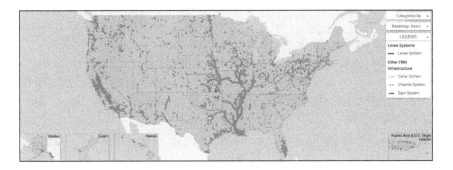

Figure 9. The US has 29,500 miles of levees, shown in black (red on the colored PDF download version), with an average age of fifty-five years. Most are not engineered for rising sea level, nor for the extreme rains and runoff now happening. Interactive map is at https:// levees.sec.usace.army.mil/#/.

The important lesson from that stunning revelation about New Orleans following the hugely expensive "fix" is that we need to begin planning realistically for this global challenge. Hoping or pretending that sea level will not rise is irresponsible. The longer we avoid facing reality, the more expensive it will ultimately be. The tough planning decisions that need to be confronted in the city known as the "Big Easy" will be anything but. The real value of this fourteen-billion-dollar project might be that it awakens all of us to the folly of shortsighted planning.

Jakarta, Indonesia's capital, tops the list of sinking cities globally. Home to some fifteen million residents and regional headquarters for many corporations, it has been struggling with the highest rate of land subsidence and SLR of any major city in the world. Recent figures show that the local rate of SLR is up to 10 inches (25 cm) a year, and totals more than 10 feet (3 meters) in the last three decades, mostly due to land subsidence. As the effective water level continues to rise, fragile levees and walls that hold back the dozen rivers running through the city have been extended even more precariously. In 2019, after decades of trying to stem the problem without success, the government announced they would move the capital to Java, another island, seeking stability, safety, and sound investment in their future. It's still not clear what that decision means for current residents of Jakarta. While the government will create a new capital on higher ground, what about the many millions who live there in poverty, who are faced with a city that could have a tragic flood at any moment?

All these places that are especially vulnerable to SLR due to extreme subsidence tug at us emotionally. They are the first wave of places under siege that will likely not survive as we have known them. That is all the more reason that we need to think forward several decades to see where water levels are headed, so that we might head for safety.

LOW-LYING ISLANDS

Typically cited as most vulnerable to disappear are Kiribati, Tuvalu, Maldives, Marshall Islands, Seychelles, Solomon Islands, and Palau. Their situations vary somewhat in terms of elevation, geologic structure, and how freshwater resources are impacted. In the media, there is often a question of which island or group will disappear first, which turns out to be a naïve, even silly question. How do you define when an island has gone underwater? Is it when the last square meter of land disappears? When the first house is fully submerged? The very last house? Half the houses? What if several houses remain intact, but there is no water to drink and no functional port or airport? Like most contests for who is first, it depends on how the question is framed. There can be multiple "winners" of such contests, though here they are all losers. Rather than being fixated on who goes underwater first, we should put the attention into how the millions of residents of thousands of low-lying islands adapt in the short term and plan for relocation in the long term.

Some rather profound legal questions will come into play. As sea level rises, properties will go permanently underwater. Where do residents go when their land or their country disappears? Is there legal recourse—as Palau is exploring? Does the country retain its vote at the UN? What about valuable fishing rights, even if it sold them and is just collecting a license fee for them?

I gave a brief overview of the legal issues in Chapter Twelve of *High Tide on Main Street*. For those interested in a more detailed view of the legal issues, I recommend attorney Margaret Peloso's 2018 monograph, listed at the back of this book under "Other Resources."

Surprisingly, the question of what is at risk is much more subjective and elusive than one would expect. For example, there is rarely any mention in the media of the Bahamas, the archipelago almost touching Florida. The Bahamas comprise dozens of inhabited islands and are home to a population of almost 400,000 people. As the four million annual visitors to the Bahamas arriving by airplane and cruise ship probably have observed, most of the country is quite low. According to a study published by Columbia University, the islands of the Bahamas are literally at the top of the list for the percentage of population at risk from rising sea level, with a staggering eighty-eight percent of their population in the danger zone.[21] The fact that the Bahamas is at the number one position is very much a function of the format and parameters of the particular study. Bangladesh and Vietnam have a hundred times the population at risk, but not quite the percentage. Kiribati and the Maldives effectively have one hundred percent exposure but did not meet the criteria of places surveyed. Like most polls and statistics, the headline ranking is largely a function of how the analysis is structured. As the saying goes, *there are lies, damn lies, and statistics.*

The entire region of the Bahamas and the Caribbean is at risk, albeit in different ways. The Bahamas is spread out over a large area, with diverse islands from Bimini to Harbor Island that are quite different from the iconic capital of Nassau. To the south, Turks and Caicos is similarly low-lying and made of porous limestone. South of that lies the actual Caribbean, which is geologically quite different. But even on islands with high elevations such as Cuba, Dominica, the Virgin Islands, and Puerto Rico, much of the development has occurred right down near sea level. Those islands with high elevation have an advantage but will want to begin changing their building codes and development planning now if they want to reap the benefits thirty years from now. It's understandable that those tourist destinations do not want to portray themselves as vulnerable and jeopardize further investments. Nonetheless, the sooner they start the process of adaptation—by changing building codes, zoning, and infrastructure—the likelier they are to be resilient and thriving in the future.

HOUSTON AND LONDON: WATER FROM BOTH SIDES

Increasingly cities will face floods from two sides, the ocean coming in on an extreme high tide or storm surge, and heavier rainfall and runoff from the inland side. During the record rains associated with Hurricane Harvey, seawater was pushing up the massive Houston ship channel toward the refineries and the city just as 40 inches of rainfall was running downhill to any low spots and accumulating to greater depths. The terrible flooding caused renewed interest in creating a storm surge barrier, which now seems likely to be built. This would be a variation on the famous harbor gates at Rotterdam covered in Chapter Five. Such a storm barrier will handle a hurricane surge coming up the channel but will do nothing about the disastrous rainfall and runoff, the cause of most of Houston's flooding. Subsidence is also adding to their challenges.

In London, there is a variation to this situation that illustrates the complexity of the challenge some locations will face. The now-iconic barrier on the Thames began operating in 1982 to protect the city from storm surge and extreme high tides. In the last few years it has been closed far more frequently; in 2013, a record fifty times.[22] The water level now exceeds its design criteria. An improved version is being planned, but there are conflicting flood challenges: to keep out the rising water coming in from the sea to the East and to deal with the water coming downstream from the midlands. With rising sea level, extreme high tides and progressively heavier rainfall, there are times when the water comes at London from both directions. The risk of the Thames overflowing its thousand-year-old banks increases. Fortunately, the local authority, the Environment Agency, is fully aware and quite sophisticated with its planning and engineering.

A LOOK AT BANGLADESH

With 160 million people in an area the size of the state of Wisconsin, Bangladesh is one of the most densely populated countries in the world. It's ranked the eleventh-poorest nation and is one of the lowest in terms of elevation above sea level. Over six percent of the country is already underwater, which will rise to approximately twenty percent with just 1 meter (3 feet) of rise.[23] That's in addition to the increasing rainfall that is already causing severe flood events. It is estimated that four million Bangladeshi have already been displaced by rising waters, a figure that is expected to increase to eighteen million by 2050. Working with Dutch engineers, Bangladesh has set up several projects to capture silt from some of the fifty-seven rivers crossing their country to create new land. Because of all the rivers, their soils are generally rich and fertile.

An amazing story can be found in Bangladesh about the threat from rising sea level and the human ingenuity response. Families on the front lines—on the rivers and in the flooding flats and marshlands—are doing what humans do best: adapting out of necessity. As their land disappears, they are resurrecting an old technique for floating farms made out of water hyacinth, bamboo, and dried cow dung. Building them entirely from natural materials, plus a few plastic jugs, costs nothing. They have expanded the concept beyond growing vegetables, using the floating farms to raise chickens and ducks, producing eggs, and farm-raising tilapia. In another example of innovation, children are picked up for school by boats that are self-contained classrooms, with computers and LED lighting powered by solar panels.[24] These wonderful examples of human ingenuity remind us of a truth: when we are pushed, we become inventive. Those who have no easy options for their families will make their own path.

BIG BARRIERS: BAY AREA, BALTIC, CHESAPEAKE

Some people in the San Francisco Bay Area have floated the concept of a Golden Gate Barrage (barrier) to isolate the bay from the Pacific Ocean. The goal would be to maintain the current shoreline and high-value real estate all around the Bay Area. At present it's little more than a crazy idea. In the other large US bay, the Chesapeake, there is also discreet talk about creating a dam or barrier to keep out the rising sea. A quick look at the globe shows at least eight seas, gulfs, or bays—semi-contained oceanic bodies—that could possibly be sequestered from the larger ocean as it rises. In addition to San Francisco Bay and the Chesapeake, the Gulf of Mexico, Hudson Bay, the Mediterranean, Aegean, and Red Seas, the Arabian Gulf, the Baltic, and the Sea of Marmara off Istanbul are potential contenders for gigantic-scale civil works projects to maintain current water level and shorelines. Whether any of them proves to be feasible from an engineering, economic, or ecological perspective remains to be determined.

RIVERS AND LAKES: EXCITING NEW FUTURES

In stark contrast to those galactic-scale engineering projects that almost defy imagination, there is one positive transformation that is safe to predict. There will be a resurgence in the practical use and economic value of most of the world's non-oceanic navigable waters and lakes. There are many nuances and variations of the term *navigable waters.* My usage implies that a cargo ship can access the river or lake from the ocean. That would include most major rivers, taking into account the locks—the systems that allow ships to change water levels, such as around waterfalls. Some large lakes, like Lake Tahoe in the US and Lake Baikal in Russia, are isolated from major ship traffic. However, the US Great Lakes are accessible from the ocean via large waterways and locks, so they would be part of the navigable waters as I am defining them.

The Great Lakes are all far above sea level. Lake Ontario is 246 feet (75 meters) above sea level; the other four are approximately 600 feet (182 meters) above. Given our earlier context of the ice sheets during the recent ice age, it's worth mentioning that the Great Lakes are very recent changes to our planet, essentially just *gouge marks* created by the glaciers. Roughly ten thousand years ago they filled with water, becoming the geologic feature they are today. Although there have been some recent modest changes in water level in the Great Lakes, that's largely from the higher rainfall associated with a warming planet.

Although there is robust shipping traffic on many of the world's rivers and large lakes, most sea freight now moves from ocean ports. In the coming decades, some of those major coastal ports will face challenges to adapt to SLR. In addition to modifying their port facilities, there will be the impact of increased flooding on rail lines and roadways to consider.

Ports on rivers and large lakes, however, will be quite stable, which will likely become a strength, or feature, with big benefits. In other words, if there is a route for shipping goods by sea from a port city on the Great Lakes or on the Mississippi, Amazon, Thames, or Elbe rivers, it will not be adversely affected when global sea level is 3 meters higher, just as an example. As a result, the communities served by those navigable waterways should have a strong future. Such cities—isolated from sea level rise but accessible to the ocean shipping lanes—may even see a very positive effect on their property values with expanded commerce.

Even aside from the economic benefits of the shipping routes, it is easy to foresee that communities on stable rivers and lakes may benefit from the growing uncertainty about oceanfront locations. Water views are prized in either setting. As a result, there is good upmarket potential for many properties on and in the vicinity of lakes and rivers. The changing climate will create important advantages and disadvantages on some large lakes and even on one shore versus another. For example, in the northern hemisphere, with the prevailing westerly winds, there is a prevailing "lake effect" on the eastern shore of most lakes that will tend to bring heavy rain or snow to that side. With warming temperatures and more moisture in the air, the precipitation may well increase over historic norms and could affect property values. All that should be considered in a vulnerability and value assessment of particular locations.

GETTING PRACTICAL: ADAPTIVE STRUCTURES

In a world that is changing fundamentally with rising sea level and changing flood patterns, the two guiding principles for all engineering, construction, and zoning need to be: (1) a margin of safety in the face of uncertainty, and (2) an ability to adjust or adapt. First and foremost, this is due to the unusual uncertainty for the outlook of thirty years and beyond. This will be a massive challenge, as our entire world has generally operated on the principle that there was consistency over long periods that could be averaged into "norms," often with predictability and probabilities.

It is easiest to think of these principles in terms of designing and building a house because we can all identify with it more intimately and intuitively, but the principles apply more broadly to commercial buildings and industrial plants. An important aspect of this is building codes and zoning, which I will cover in coming chapters. For the moment, I want to help frame some ideas about *solutions*. To do that, we should be clear on our design assumptions, parameters, or instructions.

Some planning parameters might be:

- For a thirty- to fifty-year time frame, plan for sea level up to 1 meter higher (approximately 3 feet).
- Where one-hundred-year design life is appropriate (as in most buildings and nearly all infrastructure), 2 meters (~7 feet) of rise is very possible, but not the highest possible.
- Whether higher sea level of 2 or 3 meters (7–10 feet) will occur by the year 2120 is not possible to know now. Accordingly, two approaches are useful: higher is better, and design so that further adaptation can be accommodated.
- For select supercritical structures like nuclear power plants, special rules should apply, anticipating at least 3

meters (10 feet) of SLR, with contingency adaptation engineering to be considered in the event of even higher SLR.

- In addition to increases in base sea level, allow for short-
term flood events, such as coastal storms, heavy rainfall
plus runoff, and peak tides. Recognize they can happen in
combination and that rainfall and storms could well be
higher than historical records due to the warming ocean
and disappearing polar ice cap.

All of the above should inform the "finished floor elevation" (FFE),
the key vertical threshold referenced by architects, as well as affecting the placement of electrical and mechanical equipment; drainage;
access, both routine and emergency; and utility services of water,
power, and wastewater.

While this is challenging from architectural design, engineering,
financing, and legal standpoints, some places and professionals will
solve it better than others. Those that do will benefit and profit as it
becomes clear that they have a margin of safety and have asset values
that are more secure. Big benefits can accrue for the communities and
the professionals who get in on this early and become identified as
early adopters for secure future coastal communities. Note that I did
not include insurance in that list of key fields. Surprisingly, insurance
is in a very different and less exposed situation, as we will see.

With all that in mind, here is a range of techniques and approaches
that should be considered to deal with rising water levels, both sustained and temporary, without relocating to higher ground:

- Walls, levees, and pumps can simply isolate a structure
from the sea. It requires water-resistant material such
as concrete, metal, resin, plastic, carbon fiber, clay, or
granite. That material can make it possible either to raise
the land higher or just to keep the water out, as with the
Dutch polders.

Polders are areas below sea level that have been drained to be usable land. A dike, or levee, is established around the perimeter. Then the water is pumped out. The Netherlands made this technique well known, starting a thousand years ago using windmills to do the pumping. Polders require impermeable geologic structure and thus will not work in areas with porous limestone, such as South Florida and most coral-based islands.

- Pilings and pedestals can be used as an architectural feature. This can be basic, as is common in the Florida Keys, or bold and artistic.
- Raising ground elevation with fill is fine for limited areas, but not realistic as a global universal approach. For one thing, there is simply not enough dirt to raise coastlines worldwide.
- Floating homes or houseboats are common in many waterfront areas, adjusting with the tides. Though they have obvious advantages in terms of rising with sea level and temporary flooding, some special concerns need to be recognized:
  - They need to be sheltered from storms and large waves.
  - If they are on the ocean, saltwater corrosion and maintenance can be an added challenge.
  - Hookups for water, power, and wastewater are required, though some can now operate with renewable and self-contained systems.
- Amphibious homes are somewhat similar but are normally placed on the ground and designed to float just during short-term flood events. In Maasbommel, an area of Rotterdam, a community of such homes has now been there for a decade. They move up when there is high water and maintain full utilities through a flexible bundle of wires and tubes called an umbilical. Given the great

increase in the frequency and duration of flood days,
I think there's real merit to such designs, which provide temporary elevation for hours or a few days during
floods. It's rather easy for them to accommodate 10 feet (3
meters) or more.

- *Jackup* structures borrow a method used for large off-
shore oil rigs where three or more legs are lowered to the
seabed and the structure is raised safely above the water.
As with houseboats, issues of concern are exposure to
storms and long-term maintenance needs and costs, given
the highly corrosive saltwater environment.

- RVs (recreational vehicles), motor homes, and mobile
homes are already increasingly popular, particularly in
North America, for vacation trips and increasingly as
primary homes. A range of purpose-built communities is
rapidly growing to provide them with concrete pads, utility hookups, and amenities ranging from swimming pools
to community centers and restaurants. Some of these
parks rent the spaces while others sell them like units in a
condominium. In an era of uncertain flooding these will
surely have more appeal as the easiest way to relocate and
to simply move your biggest asset—your home—out of
harm's way. For people who want to have a second home
for a change of seasons, there is a certain economic logic
to having your sizable investment on wheels, with the
option of renting or buying relatively inexpensive spaces
in the places you want to live for part of the year. The
sophistication and features of the current generation of
mobile homes are quite impressive.

- Some structures might even be built on "skids" that allow
them to be pulled uphill and away from the water when
it's necessary.

- There are now designs for floating communities, sometimes described as *seasteading.* Borges Architects have
proposed a design for platform cities in which an entire
elevated community is developed on a structure above
the water, somewhat evocative of the elevated High

Line in New York City and similar structures elsewhere. Renderings of both are shown in Figure 10 below.

Figure 10. Communities on floats and raised platforms have been proposed but are not likely scalable for tens of millions. Photo Credits: Seasteading Institute and Borges Architects

With all these inventive concepts will come essential questions of cost, durability, and scalability. The two concepts shown above are fascinating. I could well imagine such futuristic experiments being tried as proof of concept. But with tens of millions of residents vulnerable to the effects of SLR, we need to be realistic about the practical limits of such solutions.

I believe that eventually we will come to realize that no matter how much we want to hang on to the places we call home, moving to higher ground is the only sensible solution for the vast majority of people in low-elevation coastal areas. I want to be open-minded and encourage creative solutions, but it is less expensive and therefore more scalable to construct and live in residences on solid ground.

HIGH-RISE UNITS, HIGH UP, ON HIGH GROUND

As safe land becomes scarcer and more valuable in low-elevation coastal areas, residential towers will continue as the standard, just as they are now in nearly all urban environments. Their higher density, smaller land footprint, "cookie-cutter" unit designs, and factory-style construction give them the best value and lowest cost to build per thousand residents. Operating costs, security, and access to mass transportation are also more efficient. With global population already approaching eight billion and forecast to reach ten billion by midcentury, high-rise buildings would seem to be the way of the future. The concept applies to all forms of housing: upscale, moderate, and public assistance.

High-rise buildings are more cost-effective than other types of housing in terms of elevating the residential units and the mechanical and electrical equipment above the flood zone, too. An important strategic question will likely be how to balance a design that puts more such units up on even higher pilings to create a safe flood zone, versus putting the project farther inland on higher ground. The advantage of the latter is to avoid frequent flooding and the possible need to eventually abandon the building when the sea rises, compromising the coastal city. Communities will almost certainly favor buildings on higher pedestals, pilings, or fill land so that residents do not leave.

That brings up a basic difference between affected groups that needs to be recognized. Residents and business owners will likely have a reluctance to relocate, but it's an option they need to consider and almost certainly will, sooner or later. Corporations with large fixed plants or infrastructure, such as utility companies, may be unable to relocate easily. Local governance units such as cities and counties are in a different situation, as they generally do not have the option to relocate.

ARCHITECTS AND ENGINEERS SHOULD BE IN THE LEAD

It is imperative that we consider the process and the professionals who shape what is referred to as the "built environment." Architects, various types of engineers, planners, and landscape architects will all be on the front lines of adaptation to rising seas.

"Built environment" broadly refers to the creation of buildings and infrastructure, in contrast to the natural environment. It encompasses all aspects: planning, design, engineering, construction, infrastructure, and landscaping.

Architects have a special seat at the table because they look at projects from at least three perspectives: functionality, appearance, and safety. That combines the "left brain–right brain" methodologies, the logical with the artistic. The challenge of inexorable sea level rise requires that for sure. Some architects have advanced innovative designs and put the topic in front of their peers, but generally they are limited by constraints beyond their direct control, like building codes and, most importantly, the client's direction. In many coastal areas, there is a growing awareness of the issue among architects, but typically it is not front and center in their work. Individually several architects have found ways to implement higher levels of flood resiliency on a case-by-case basis with clients. In Florida, for example, their state professional organization has been discussing a forceful push for a height increase in building codes and a collaboration with other professions and state officials responsible for building standards to deal with the long-term challenge of rising sea level for several years. It's understood to be a big problem, but tackling it has been repeatedly deferred.

A NEW HARBOR CITY FAR INSIDE GERMANY

Landscape architecture is one profession that will be particularly important. One of my favorite examples is a newly developed area adjacent to historic Hamburg called HafenCity, which illustrates how inventive planning and landscape architecture can integrate with building designs to reduce flooding.

Two decades ago, thriving Hamburg needed to expand. Their special challenge is that the unpredictable and sometimes violent North Sea will often push storm surges far up the Elbe River to the major port nearly 70 miles (110 kilometers) inland. For a thousand years, Hamburg has been one of the largest ports in Europe, partly due to its safe location. Routine high tides move up and down 10 feet, but storm surges often add another 10 feet of water to that. The old port city solved the problem with an imposing fortress dike. It kept historic Hamburg functioning, but as they built the wall higher, it impeded the view of the river that was a feature of the city. An old industrial area just outside the dike was an obvious candidate for the expansion. Rather than duplicate the steep walls of the old Hamburg-style dike, the new design of HafenCity features many stepped surfaces and public areas that are actually *designed* to flood. The city raised the base ground elevation; set separate elevation heights for residential, dining/ retail, and emergency corridors; and established different engineering criteria that take short-term flooding into account. For instance, all commercial areas, including restaurants and plazas, are designed to withstand short-term flooding with watertight flood doors for the storm-driven floods that recede within hours.

Figure 11. HafenCity (Hamburg) has created a new city that is flood-resilient to 30 feet (9.5 meters) of water. Photo Credit: HafenCity Press Office

HafenCity's approach is not a solution for all places. It would not be possible, for example, to raise all of Florida or Vietnam the same way that this half square mile of Hamburg was elevated. They had a big advantage in that they could start with a clean slate rather than changing something that existed. Nonetheless, HafenCity provides a great functional example of innovative design anticipating future flooding.

ANNAPOLIS: ANCHORS AWEIGH

Just a half hour from Washington, DC, Annapolis is best known for its charming colonial architecture, its claim as America's sailing capital, and the adjacent US Naval Academy. They also have a serious flooding problem from all five of my flood factors: storms, rain, runoff, extreme tides, and sea level rise. Founded in 1649, the city is not only living history; it provides good records of how things have changed. A key traffic circle at the base of Main Street in Annapolis is very vulnerable to flooding. Fifty years ago, it typically had to be closed to traffic four times a year due to deep water. By 2016, it was being closed about forty times a year, a tenfold increase. The damaging impact of the routine flooding on residents, tourism, and the business community is unmistakable . With over a thousand registered historic buildings, the city faces a special challenge. Altering, elevating, or relocating any of those centuries-old homes and landmarks is not to be taken lightly, as it impacts their authenticity. The sea, the dominant feature of this seafaring community, now threatens it.

In 2016, the mayor's office and the office of historic preservation had me give a series of presentations to their staff, the business community, and the public to help frame the big view of the SLR problem and where things were headed. On a cold winter night, the large theater at St. John's College was surprisingly filled to capacity. There were some good follow-up meetings and workshops. But in the end, there was no strong plan for the future, at least not on the scale I had hoped to see. I pushed for a thirty-year plan, but that never happened. When the highly visible and frequently flooded City Dock parking lot had to be redone a few years ago, it was rebuilt just a couple of inches higher. That would have been the perfect time to redesign for the future, raising it at least 18 inches. Their reasons for not doing so included budgets, concern for the historic look, and some state regulation related to the public works budget. So the flooding continues to worsen. Annapolis is one of my favorite cities. Hopefully they will find a way to balance their history with planning for the future. (Perhaps, to put it in sailor-speak, *it's time to loosen the lines and rise with the tide.*)

During my visits to Annapolis, I was introduced to several department heads at the US Naval Academy. That led to the great privilege of giving a featured lecture there. "Rising Sea Level: Engineering Challenge of the Century" was attended by hundreds of midshipmen and a large number of faculty and was well received. I learned that the academy has a very intimate awareness of the problem. Classrooms on the ground floor now flood at extreme high tides and in storm conditions. At what is arguably the world's premier institution for ocean engineering, future navy officers have been seen taking off their shoes and rolling up their pant legs to keep their uniforms dry during extreme king tides.

After my presentation, Professor David Kriebel, chairman of the Ocean Engineering Department pointed from his office window across Dorsey Creek to the sacred officers' cemetery and the *columbarium*—the structure for burial urns—built in the early 1990s right down near the water's edge. He told me that in recent years, rising sea level floods the area with increasing frequency at peak tides, even under clear skies. It seemed really poignant that they now have to check the tide table to schedule burial ceremonies so the creek won't be washing over guests' ankles. With 20/20 hindsight, they should have built it higher. But three decades ago, rapidly rising sea level was underestimated even at the US Naval Academy. A robust, aesthetic, and durable solution is needed for the challenge posed by rising seas at the iconic institution. Given its wonderful architecture, putting in steel sheet piling and pumps seem less than ideal.

Moving the campus to higher ground seems unimaginable, at least for the foreseeable future. Perhaps they will raise the lower buildings. While there is no easy solution, a forward-looking solution for the academy might serve as a model for the larger challenge of how the world will reengineer itself in the face of the rising sea.

7

INSURANCE AND PUBLIC POLICY MAY INCREASE RISK

You can never plan the future by the past.
—Edmund Burke

Insurance and government action are widely seen as solutions to flood risk. Unfortunately, both have a time horizon that not only can mislead, but also encourage bad planning and policy in the face of an ever-rising sea. We need to look at how misunderstandings about risk, insurance, and public policy may actually be making the long-term scenario for SLR disaster much worse. Unstoppable and accelerating rising sea level requires that we see where things are headed so that private and public policies are realistic and create good incentives to reduce risk and invest in assets that will have durable value.

Nothing much is built or financed without reasonable confidence about the security of the assets over their useful life. Assets here can refer to homes, commercial buildings, industrial plants, communities, corporations, and infrastructure—including roads, railroads, bridges, tunnels, ports, and airports. The durable and useful lifespan of assets varies considerably. The period for which they are financed gives a conservative indicator of asset life. But from homes to large commercial buildings to roads, railroads, and industrial plants, it is common for us to use and benefit from assets far longer than the finance period, often for more than one hundred years. It is this new duality of flooding, the short-term catastrophic events, and permanent sea level creeping slowly higher, that is so challenging when it comes to figuring out how to calculate, assign, and reduce risk. Our entire experience and system are designed to address the short-duration flood events only.

Wrongly confusing quick flood-disaster events with the unstoppable rising sea is setting the stage for a whole new level of catastrophe. Devastating, deadly floods are nothing new; they can be traced back to biblical times. In the modern era, however, dramatic population growth and increasing density are rapidly increasing the exposure on the coasts as well as on inland floodplains. In addition, there are now the added factors of extreme storms, deluge rains, and rising seas. Judging risk by the usual hundred-year or even five-hundred-year record-level events is no longer a good guide in this new era.

The unprecedented and potential exponential growth in SLR warrants a new view of government policies and insurance structure, so that what is done in the near future helps build a solid foundation for midcentury and beyond. Government programs that promise affordable flood insurance, or that help citizens rebuild or be bought out of their unusable properties clearly have appeal, but need to be evaluated carefully in order not to encourage more risk and exposure. Whatever we do now creates expectation and entitlement, if not legal precedent.

Paradoxically, however, the usual insurance time horizon actually hides the risk of accelerating SLR, in many ways *increasing* future risk. This point is critical to our understanding of the problem and charting a path forward.

At its simplest, risk is based on the probability of something happening. Actuaries are the professionals who determine the percentage risk of something occurring, which can then be priced as an insurance premium by someone who actually underwrites or takes on the risk of the insurance policy, which is a contract. To make this actuarial process clear, let's take a realistic but highly simplified example of fire insurance. If records show that wooden houses in a certain area and climate have a one in three hundred chance of being destroyed by fire in the next calendar year, it's quite easy to put a price on the probable fire risk of a typical house in the property profile. That enables an insurer to offer a one-year insurance policy that would be profitable and competitive in the market. It is all based on being able to calculate and average the risk over thousands of individuals over several years. The fundamental assumption is that the probability or "odds" of future events occurring can be found in decades past. The ability to pool and price all manner of risks makes our modern world of cars, airplanes, massive complex buildings, professional practice and liability, and all manner of industry possible.

Basic risk sharing goes back as early as four thousand years to the Code of Hammurabi, but what we think of as contractual insurance policies began in Genoa in the fourteenth century. Insurance concepts advanced further in the aftermath of the 1666 Great Fire of London. An innovative market model to insure the high risk of ships and cargo started in Edward Lloyd's coffee shop two decades later, operating today as the iconic Lloyd's of London. With its unique structure as a marketplace with nearly one hundred separate underwriting syndicates, Lloyd's fosters innovative insurance products.

From ancient times up to the present day, large flood events posed a special problem, however. Whether it was a hurricane or record rainfall causing a river to overflow its banks, floods could destroy an entire region at once, involving tens of thousands of homes and businesses, which could bankrupt the insurance company responsible for providing compensation on all of those properties. In principle that problem led to the creation of the US National Flood Insurance Program in 1963. Policymakers believed that only the federal government had the capacity to cover such risk. Also it would be possible to establish better policies for where and how to build communities. We will look at that United States program in a moment, but first need to understand the concepts, since other countries handle flood risk quite differently. In the United States and more generally, as flooding gets worse by the decade, four priorities are widely seen as part of the solution:

- Affordable flood insurance so that people have coverage
- Improved flood risk maps and models
- Disaster relief funding when there is a disaster
- "Buyout" programs for public purchase of properties at risk, usually at full valuation, as a path to their removal

The problem is that all four of these well-intentioned efforts can make the long-term problems associated with sustained rising sea level much worse. If we are going to adapt smartly and minimize the coming catastrophe, it's essential that we understand the pitfalls. The deadly destruction of flooding presents one of the most fundamental and primal risks to property as well as to people.

Two things confuse or obscure our understanding of flood risk, even among experts. First is the confusion of flood types. You may recall the very different *five flood factors* I described in Chapter Five. Whether you are the property owner, the banker, or the insurer, you will not be able to project the risk of future flooding if you do not differentiate between the forces and frequency of storms, rainfall, downhill and downstream runoff, extreme tides, and sea level rise. Even something as subtle as the nineteen-year lunar tide cycle (Chapter Three) has to be understood. Imagine you evaluated a critical flood height in a tidal area for the five-year period 2018–2023 and found little if any rise. If you did not understand that that was during the down period of the nineteen-year repeating pattern, you might conclude that SLR was negligible. You would likely underestimate floodwater heights that will start increasing in 2024 as we enter the next up-cycle phase of that planetary cycle affecting tide heights. Increasing rainfall patterns also have to be seen as a changing component of future flooding. (Coastal erosion may aggravate coastal flooding but needs to be looked at separately for analysis and future forecasting.)

The second cause of fundamental confusion about flood risk is the result of very different timescales. For one thing, SLR is totally unlike the other four flood factors and trends. As an element of flood risk in the next twelve months for an insurance policy, higher sea level at the current rate of a quarter inch per year (5 mm) is so small as to be irrelevant. The possibility of enough ice entering the ocean from Greenland or Antarctica during the annual policy period to raise sea level 1 foot is nearly zero. The possibility of breaking flood records due to a combination of the other four flood factors is essentially the only thing the underwriter needs to assess for that one-year period.

Our understanding of vulnerability and our decision about moving to higher ground can only be made intelligently if timescales are aligned, which is not at all the case with insurance and the general public discussion. For that matter, the long-term risk is misunderstood by many in the insurance industry. As we saw in Chapter Five, an engineer may look at the very same project with time horizons of one year, thirty years, or even a hundred years, depending on the client and purpose. In my example with the bridge, the short time horizon was for an insurance underwriter. Insurance is well understood to be a financial contract based on the risk or probability of something occurring—or not. There are many forms and layers of insurance. In the area of property and casualty insurance, it is almost always done on a year-to-year basis. There are trillions of dollars of insurance contracts, but they all follow the same principle: For a very specific period, if you pay us a certain fee, we will pay you "x" if "y" occurs. Arguably, it is a legalized form of gambling.

Insurance here refers to the actual **underwriting** or financial responsibility for the risk, not the more visible firms that sell and service the policies at the retail level. Underwriters are the ones who are responsible for the risk and who pay out in the event that an insured event occurs, and who profit in the absence of that event. A small number of firms are categorized as **reinsurers,** who do a second level of underwriting, effectively selling an insurance policy to an insurance underwriter to further spread the risk. All those may be grouped as insurance underwriters for our purpose here.

Anyone interested in building or buying property will want to know how safe and secure it is. The owner might have an engineering assessment done, but often the risk is assumed to be represented by the cost of the insurance policy. Indeed, any competent insurer will have had an engineering evaluation done to assess the security and the risk. If you borrow money from a bank or look to others for financing, they too will be very focused on the security of the assets and the soundness of the various insurance coverages. With regard to changing sea level and flooding, a fundamental mismatch of timescales is inhibiting good decision-making and setting the scene for epic disaster.

In a classic book on risk, *Against the Gods: The Remarkable Story of Risk*, Peter Bernstein[25] explains how our civilization is largely predicated on the ability to assess and value risk. He poses a critical issue that has real relevance to our current challenge of planning for the glacial meltdown and rising sea level:

> *The issue [of risk] boils down to one's view about the extent to which the past determines the future . . . To what degree should we rely on the patterns of the past to tell us what the future will be like? . . . The mathematically driven apparatus of modern risk management contains the seeds of . . . self-destructive technology.*

He quoted Nobel laureate Kenneth Arrow's warning: "Knowledge of the way things worked, in society or in nature, comes trailing clouds of vagueness. Vast ills have followed a belief in certainty." He points out that the insurance industry and its clients have been going down a path of more and more competitiveness and sophistication of product, but perhaps falsely assuming that the past predicts the future. The problem is that in today's world, the average flood frequency and climate conditions of the last five hundred years, hundred years, or even thirty years gives us little if any information about the future. In fact, our training, instinct, and products like insurance are largely misleading, since their assumptions about the future are based on the past.

The implications of a fundamental sea change in risk have been recognized at a high level in this sector. A nonprofit think tank for the Swiss insurance industry, the Geneva Association, issued a short but profound and very relevant report in 2013, *Warming of the Oceans and Implications for the (Re)insurance Industry*.[26] This internal document for the insurance giants makes clear that the assessment of future risk should not be based on the past, making these four key points (paraphrased):

- Robust evidence that the global oceans have warmed significantly has effectively caused a shift towards a "new normal" for several insurance-relevant hazards. Warmer oceans will change global weather in the short term and sea level in the long term.
- This shift is quasi-irreversible. Even if greenhouse gas (GHG) emissions completely stop "tomorrow," oceanic temperatures will continue to rise, far into the future.
- In the fundamentally changing ("non-stationary") environment caused by ocean warming, traditional approaches, which are solely based on analyzing historical data, increasingly fail to estimate today's hazard probabilities. A paradigm shift from historic to predictive risk assessment methods is necessary.
- Due to the limits of predictability and scientific understanding of extreme events in a non-stationary environment, today's likelihood of extreme events is ambiguous. The implications are profound.

Note that this insurance industry think tank has focused on the same concerns as the US Department of Defense report cited in Chapter Three. They are both saying that future base sea level and flood events will surely be quite unlike those of the past, which runs directly counter to the way we typically look at risk. Both realize the limitations of our ability to predict precisely, and both industries encourage serious consideration of extreme scenarios, particularly with regard to future sea level.

ONE YEAR VERSUS THIRTY REVERSES SLR THREAT PRIORITY

Insurance underwriters issuing one-year policies actually have little concern about sea level rise even a decade from now. A hypothetical example may help. Say you were looking to issue a flood insurance policy for a home in Charleston, South Carolina, that was exactly 10 feet above sea level. A good vulnerability assessment would include the worst-case scenario: momentary flooding that could happen during the one-year policy due to some combination of coastal storms, peak lunar high tides, heavy rains causing direct and indirect flooding (from uphill sources), river flooding, and land subsidence. The analysis might determine that there was a three percent chance that water level would exceed the 10-foot (120-inch) flood threshold during the one-year period of the policy. Note that I did not include flooding from global SLR. The possible additional water level specifically from rising global average sea level during the next twelve months is negligible, perhaps a half inch in the extreme, so quite irrelevant to the risk of flooding that property within the yearlong policy.

Now let's consider the exact same situation thirty years from now. That assessment for the same property might add a few more inches to the flood probability assessment due to heavier rain, subsidence, and worse storm patterns, but for this example, let's assume the vulnerability assessment does not change for any of those factors cited just above. Then there is sea level rise. Assume that global average sea level rises 18 inches (47 cm) over the three decades, in accord with some of the worst-case projections. You can easily see that the higher base sea level height would have an enormous effect on the probability of flooding and on the calculated insurance premium. Instead of taking 10 feet of water to cause flooding into the home, it would only take 8.5 feet of short-duration flooding. While it is hard to say what the flood insurance market will be like thirty years from now, in a scenario with sea level being 18 inches higher, experts suggest that flood insurance premiums could easily double, assuming it was even available for that location. If the cost of flood insurance doubles, it reduces the desirability and value of the property significantly.

The point is that the current cost of flood insurance does not reflect the full risk of future flooding or submergence over the next few decades of its financing, let alone for the life of the asset. The lack of any regulatory or market mechanism to look beyond the current single-year insurance premium as a risk valuation is a huge problem. It has tremendous implications for owners and financiers of coastal assets, as well as on all the enterprises that are dependent on those assets.

In 2018, *Actuarial Review Magazine* interviewed me for a cover story titled, "The SLR Factor—As Sea Levels Rise, the Flood Risk Equation Changes."[27] Again, actuaries are the specialists who calculate the risks and premiums according to probabilities. The article quoted another expert, Dag Lohmann, the head of KatRisk, a catastrophe modeling firm, who said, "Along the coasts of the United States, there is about 6.88 trillion dollars in exposure potential [to flooding]." I was intrigued by the specificity of the number, so I contacted him. He told me it was largely an estimate for storm surge, and that they did 50,000 years of model runs in a Monte Carlo simulation. Fair enough; sounded comprehensive. Then I asked how high sea level rise was in the model that yielded the $6.88 trillion of potential damages. He said it really did not matter. At first that stunned me, since the property value at risk would surely change if sea level was several feet higher. Then I thought about it from the view of the underwriters—his customers—and finally understood why SLR was largely immaterial to the insurance professionals. From the insurer's standpoint at present, the curves for future flood risk *beyond* a few years are largely irrelevant.

That myopia, or shortsightedness, is not limited to the insurance sector. Engineers and modelers may do something very similar, but for somewhat different reasons. For example, DHI, with headquarters just outside of Copenhagen, has an excellent global reputation for sophisticated water modeling. At their office in 2017 they showed me their latest flood model, with granular detail of flooding in various parts of Denmark based on different situations of storms and rain and adding future sea level rise on a dynamic basis. It was quite impressive and no doubt useful for civil engineering, flood preparation, and emergency management. They told me they ran it at different heights for SLR according to the IPCC report, with the maximum being the 90 centimeters (32 inches) covered in Chapter Three.

I asked if they had ever run it at two meters of SLR but was told that no client had ever asked for that. I pointed out that Antarctic melting was essentially left out of those IPCC projections for sea level but could easily add another meter or more. They were challenged by that information and agreed they should revisit the range of scenarios for which they do models. It reinforced my concern that even experts looking at future flooding are compartmentalized, not realizing the limitations or omissions in the top-line sea level rise numbers in the IPCC report and therefore not preparing accurately for the new reality.

The Dutch consulate in Miami had set up a meeting for me in The Hague later that same week, with a senior official in the Ministry of Infrastructure and Water Management. He was the project manager for the Global Centre of Excellence on Climate Adaptation, which the Netherlands was creating for the United Nations. I asked how much sea level rise they were planning for this century as a worst case. Ninety centimeters was the response, *per the IPCC*, virtually the same answer I had heard days earlier in Denmark. When I asked how they would handle the possibility of two meters of SLR, he shook his head slowly and said, "We couldn't cope with that." Again I explained how Antarctic glacier contribution was not even in the projections for that 90 centimeters of higher sea level in the fourth and highest IPCC scenario. Clearly disturbed, he said, "So, I guess you're concerned about the new iceberg breaking off from Antarctica raising sea level." He was referring to the gigantic iceberg "A-68" that was in the news as imminently breaking off from the Larsen ice shelf. It was clear that he shared the widely held misunderstanding that melting icebergs affect sea level. I explained the phenomenon that floating ice shelves and icebergs do not directly add to SLR as they break up and melt. He quickly grasped that it was when the ice on Antarctica moved into the sea, as new icebergs or meltwater, that sea level was affected. Thanking me for taking the time to explain, he said he was fairly sure that most people had the same misunderstanding. More recently, in 2020, I had a similar experience with HR Wallingford, an advanced coastal modeling agency in the United Kingdom, where they too used the IPCC's highest projection for SLR this century, 90 centimeters, as their worst-case scenario.

My point is not to impugn any of these fine organizations or the people involved, but to illustrate our difficulty with planning even when the geologic record, clear physics, and current observations all point to the very real possibility of dramatic sea level increases ahead largely due to processes now underway in Greenland and Antarctica. These three examples, along with many other encounters, have made clear that even experts concerned and involved with climate change can have blind spots. I think there are several reasons:

- Accelerating information flow: In this information age, there is so much information that no one can cover it all. Everyone becomes specialized, even compartmentalized. Even if someone is working on sophisticated models or a program aimed at better education and policy, there is no reason to expect them to have a holistic understanding of climate change.
- Weak assessment of risks: Engineering, financial, and risk assessments are made based upon assumptions that often are not scrutinized.
- IPCC projections are scientifically conservative: The way that the IPCC states the projections for SLR without drawing attention to the misleading lack of Antarctic contribution to the projections is causing organizations who rely on that data to plan inaccurately for future scenarios.
- Alleged solutions may not be vetted: Wildly impractical academic or theoretical solutions tend to go unchallenged.
- Misunderstanding of CO_2 reduction strategies: There remains widespread confusion over the fact that reduction of carbon dioxide emissions cannot stop SLR this century.
- Human realities: Engineers are human too. At some level, many of them do not want to consider that sea level rise is unstoppable, and that it could get 2 meters higher, or more. The implications are simply too disturbing and result in cognitive dissonance.

FLOOD INSURANCE IS NOT WHAT IT SEEMS

With the context that sea level is rising and accelerating, adding to the other short-duration flood events, flood insurance is widely presumed to be a key part of the solution. Flood insurance varies greatly by country. There is no way to cover the diversity of programs here other than a few examples to understand where the trend of future increased flooding will intersect property values. Before we get to the extreme version of flood insurance, the US National Flood Insurance Program (NFIP), it's worth getting a range of perspectives using the Netherlands and Canada as examples.

It might surprise you to learn that in the Netherlands there is no flood insurance as a matter of public policy. Issues of flood protection are wrapped up in the larger government program of flood defenses, overseen by the Directorate-General for Public Works and Water Management, *Rijkswaterstaat*. Because of the very large scale of the unique Dutch situation where the levees, pumps, and man-made barriers are responsible for the very existence of large areas of property in the first place, there is good rationale to their policy. With entire communities vulnerable to being put underwater, all matters of coastal defense are part of national security, quite literally. Individual private insurance would pit them against the state and the collective sense of dependency. It is a solid strategic approach even if it might seem that they are "naked."

Before 2015, Canada was the only country in the G7 that did not offer actual flood insurance, which they refer to as *overland flood insurance*. Up until then, flood insurance in Canada was largely restricted to addressing damage resulting from water backing up through overburdened storm sewer systems into the basements of homes. This insurance product (sewer backup insurance) did not apply to damage from overland flooding. For example, if a local stream overflowed, causing floodwaters to enter a home through a basement window, this would be deemed overland flooding that was outside of backup sewer coverage. In recent years, as Canada has experienced greater flooding in coastal areas and inland from greater rain, property owners put pressure on Canada's property and casualty insurers to offer overland flood coverage, which they did. As of 2019, about thirty-five percent of homes eligible for overland flood insurance across Canada have opted to purchase it.

As might be expected, the most ambitious flood insurance program is in the United States. Despite good intentions, it now has created a system of perverse incentives that make little sense. The US National Flood Insurance Program is generally understood to be essentially bankrupt, often wryly described as being "underwater." There are calls for reform from many sectors, even from within the Federal Emergency Management Agency (FEMA), where the program is based. The program dates back nearly a century when there were several truly devastating floods, both from coastal storms and from inland rivers that overflowed their banks, destroying thousands of homes and taking many lives. Since floods are so costly, there was doubt about the ability of private insurers to pay the necessary sums and remain solvent. Congress saw the need to stabilize the market and establish policies to reduce exposure to loss in the future. Eventually that culminated in the NFIP in 1968. Today, roughly two-thirds of home flood insurance in America is covered by the NFIP.

Over the years, the mandate from Congress has evolved, but essentially there are three elements that may seem compatible but are often in conflict:

- Make affordable flood insurance available to American homeowners
- Provide reliable, financially self-sustaining coverage for the long term
- Encourage better resiliency and adaptation by using rules and rates to reduce future loss and damage (e.g., improved building codes, zoning, and drainage)

The operating losses are generally blamed on lenient rules that enable the program to continue insuring and paying out claims on a relatively small number of highly vulnerable properties, something that would not be tolerated with commercial insurance underwriters. However, the problem goes beyond that, I believe. In every other field of insurance, such as fire and car insurance, the industry has real exposure and therefore great incentive to balance risk and fees. With flooding, the federal government underwrites the risk, and "Uncle Sam" is left with the deficit when it comes time to pay. When Congress gives more money to bail out the program, the cost is shared by all taxpayers. The system is about as far as one could get from the normal concept of insurance. Rules of exposure are modified to satisfy the appearance that the government is dealing with the tragedy of flood loss, particularly with major disasters like hurricanes and the deluge rain that causes rivers to overflow, yet the cost of insurance—the premiums—must be kept affordable both to satisfy the taxpaying property owners (voters) and to be affordable enough that people will buy it in the first place. To get people to buy the insurance, the NFIP has marketing and sales programs handled by the commercial insurers. There are literally thousands of them, including most of the popular brand names we would recognize. The NFIP even has a nice, friendly web page under the banner "FloodSmart," where it lists the many providers by state. Yet this masks a fundamental problem. As a federal program, the NFIP has none of the usual need to balance income and costs. The federal employees managing the program are in a very different situation than they would be if they were working for a private company. Having met with some of them, I believe they are smart, responsible, and concerned. But they are caught between providing a public service that requires the support of Congress and being privy to the ominous trends that threaten to make the program's deficit even worse. This is in sharp contrast to normal private sector insurance, where it is understood that a business must make a profit to stay in the game.

As flooding from storms, rain, extreme tides, and sea level rise gets worse, the cost to the taxpayers becomes greater. Recently, the program took some modest steps to use commercial reinsurance to spread some of the risk, part of an effort to make the operation more market-based. Nonetheless, there is huge resistance to anything that increases costs or restricts benefits. In recent years, Congress has repeatedly tried to get the NFIP out of deficit by raising rates and restricting rules for what they will insure, but the political outcry quickly shredded those efforts.

THE MORAL HAZARD

From my view, the biggest problem with subsidized insurance stems from what is termed a moral hazard in economics. Our desire to do the right thing—providing reasonably priced insurance, buying out vulnerable properties, and compensating for disasters—actually *encourages* more risky behavior. In other words, providing flood insurance at below the cost that actuarial analysis suggests is consistent with the real risk encourages more building in zones vulnerable to flooding. As sea level rises, the problem will get worse—much worse. We have unwittingly perverted the market forces that would normally have dissuaded people and companies from investing in high-risk zones. You don't have to be a mathematician to see the problem with insuring millions of high-risk properties, with average values upward of a quarter million dollars, based on the perception that the government will absorb the full risk "forever."

The other part of FEMA—the part that provides disaster relief when there is a major flood—also appears to reduce the risk of building in flood-prone areas. You may have a home or property in a low-lying, flood-prone location because you can get below-market flood insurance, or you might choose to take your chances, counting on the government to come to your assistance if there is a disaster. Knowing that Uncle Sam will bail you out in a crisis can undermine the need to participate in the flood insurance program.

Another variation on the moral hazard question are the programs to buy out homeowners at market value. This was done in New York after Hurricane Sandy and is being done from Alaska to Louisiana to Virginia as routine flooding gets worse. On the face of it, this practice seems to make sense. The intent is that if it is good public policy to move away from vulnerable coastal areas, the public should buy out those homeowners. However, inadvertently that also sends the message that the government will assume the responsibility. Buying out the homeowners at "full value" transfers the risk of building on the shore to the government. Again, it becomes precedent and encourages more home buyers to take risks that they otherwise might not. Given the millions of homes that are in harm's way, how far can that be scaled?

To conclude, government policies regarding flood insurance and disaster relief need to be revised with an eye toward solvency and smart incentives and disincentives over at least a generation, not just an elected term of office. I believe that thirty years—the duration of a common mortgage, a little more than a generation—is the right planning time horizon.

Looking at the United States, the 2012 legislation to reform the US NFIP was a big step in the right direction, but was rescinded the following year due to an outcry from voters. Excellent recommendations for reform also have been made by the National Resources Defense Council and the Union of Concerned Scientists. FEMA's own inspector general made smart recommendations. But to date, the political will to fix the program is not evident. Looking more widely to any coastal country, I believe that competitive, private insurance markets should be part of the solution. However, we need new mechanisms and time horizons. Perhaps requiring that those offering flood insurance also contribute to a carefully protected fund to cover the longer term, increasing risk—sometimes referred to as the tail risk—might be one solution.

A good approach is to clearly state that government supported programs will be phased out over a few decades, giving people time to adapt. In an extreme case that might be a matter of realizing that a property owner only has until a certain year to use it.

Flood Re in the U.K. is an excellent example of such a limited period program to encourage adjustment over a modest period of years. Participating insurance companies and their policy holders get a government subsidy for a limited time. For such programs to work, it is essential that the deadline be maintained. The British government backed program was established by law with a 25 year duration, ending in 2039. https://www.floodre.co.uk/our-future/

To be sure, there is no easy answer to this, but it's important to recognize that the longer unbalanced subsidized insurance programs continue, the more assets that will be exposed and the harder it will be to change course. The longer non market-based policies persist, the more they will serve as precedent for increased risk and exposure.

8

WHEN PROPERTY VALUES REALLY GO UNDERWATER

The pessimist complains about the wind;
the optimist expects it to change;
the realist adjusts the sails.
—William Arthur Ward

Real property—land, buildings, and infrastructure—is closely linked to financial assets such as mortgages, bonds, and REITs (real estate investment trusts). However, in the case of long-term rising sea level, there are some critical differences between hard assets and financial instruments, a difference that may well have disastrous financial effect much sooner than anyone is expecting. It's important to understand how they are connected, and how they are not.

As in any open market, real estate and financial instruments of all sorts go through ups and downs in valuations. Typically, financial markets are much more volatile than real estate. High-quality land and buildings have been seen as one of the very safest and most stable investments. Of course, property is commonly financed and can also serve as security for various kinds of financing. The point is that in today's world of highly sophisticated and leveraged financial instruments, there is an interlocking aspect of property values and finance, though one that even experts find hard to evaluate fully. The ignorance of its complexity was demonstrated by the 2008 global financial crisis. The string of defaults, bankruptcies, and government bailouts started with the 2007 collapse of a single REIT, New Century. Like a falling domino, that began a two-year long chain reaction, quickly causing financial devastation around the world. The "house of cards" came crashing down. That stands as testament to how the world can be fooled if an underlying "fact" turns out to be false or seriously misunderstood. Back then the underlying issue was billions of dollars of second-rate ("subprime") mortgages, thought to be safely insured and rated. We came to learn about arcane financing tools, like collateralized debt obligations, which could far exceed the value of the underlying asset because of the overlapping, layered, and interlocking obscure contracts, "derivatives," insurance, collateral, and bonds. The financial community had come to believe that it was built upon the secure world of real estate value, which turned out to be an illusion.

A key to the recovery was the knowledge that new money would come into crashed markets sooner or later. Few questioned that the US stock market and real estate would recover in value. Buying during crashes has often been how future fortunes are made. In fact, many savvy investors think that the worse the devaluation, the better the opportunity to make a huge profit. This self-correcting phenomenon is one of the great aspects of a free market. An overvaluation bubble normally will lead to a devaluation, which invites strategic investors to buy in low, knowing fundamentals have to recover. You just had to believe in the fundamental strength of the economy and eventual recovery. Rising sea level will be different in a few important ways.

COASTAL PROPERTIES CAN GO UNDERWATER THREE WAYS

People who own threatened coastal property often ask how long they have before a location goes underwater, referring to when it will lose value and be uninhabitable. While one may think the value of a property disappears *when it is submerged* by rising ocean level, that's incorrect. Sustained submergence should actually be considered the last of three stages in which property can go down in value from flooding.

The first stage of a property going "underwater" is in the financial or figurative sense. Even before a coastal property physically experiences frequent flooding, it can go down in value as it becomes perceived as an at-risk investment due to its general position in relation to the water. It is well known that markets are quick to take future risk into account. As previously covered, in areas subject to tidal flooding, the water level has been increasing due to rising sea level these past few decades. In this first stage, the *king tide* high-water peak might stop inches from your property. But the trend, as the water gets closer over the years, is clear. That risk of likely future flooding will already be impacting that property's value.

In the second stage, the property will begin to *actually* flood more frequently due to routine king tides and sporadic storms, which will move ever higher as a result of rising sea level. A property will be devalued if it floods regularly, even once a year. The more often it floods, the less desirable and the greater the effective discount. The third stage, *going underwater*, describes a degree of permanent submergence. Those three variations are why I often say that as sea level rises, vulnerable coastal properties can go underwater permanently, occasionally, and *anticipatorily*—an unusual word construct, but accurate.

The key question, of course, is how soon that anticipation-discounting of property values could start. In some places it already has. The effect can be seen most easily in the communities where there is land subsidence, causing sea level to rise much faster than the global average. Norfolk, Virginia Beach, and the other communities in and around the Hampton Roads area are a good example. Compared to the global average sea level rise of 8 to 10 inches (20 cm) in the last century, flooding there has been about four times greater, exceeding 3 feet in some places. The routine flooding at the king tides, even in blue-sky, sunny weather, is obvious to all. For example, the local Unitarian Universalist church changes scheduled services during the expected king tide days because the lower pews predictably flood, something that did not happen a few decades ago.

If you were looking to buy a house and planning to stay there for thirty years, the fact that the neighborhood floods more each decade would very likely affect what you were willing to pay. That is, it's not just the current frequency of flooding one has to consider but also future flooding and future value. Many homes in Hampton Roads and up into the lower Chesapeake have been elevated anywhere from 4 to 10 feet. Some alterations have been funded by a government program (HUD or FEMA), others by insurance, and still others by the home-owner. Almost certainly, the practice will increase. In the Eastern US and Gulf Coast, you can shop online for companies offering competing services to elevate a house. There's even a trade association for contrac-tors exclusively doing this work.

We will not try to cover here the full range of effects on a prop-erty as the water level rises. Vulnerability assessments need to be done for specific areas, even for specific parcels. Having participated in or advised quite a number of them, I would encourage you to have one done to cover your interests if this book leaves you feeling that you need one. The point is that different people will have different interests and different priorities. Even significant land elevation may not mean the value is safe. For example, a homeowner on Long Island, a hundred miles east of Manhattan, told me his home was 80 feet above sea level on a bluff. After reading *High Tide on Main Street,* he realized that being the highest house in the area could turn his property into a very small island with little value.

It's also important to keep in mind that, in addition to coastal properties, all homes connected to daily tidal change through riv-ers and marshes are affected. Throughout human history, flooding from rivers has been a major threat to lives, homes, and crops. A lot of effort and engineering have been directed to reducing flood haz-ard. Understandably, the engineering was typically done based on a 100-year, or perhaps 500-year, flood history. Unintentionally, this has added to our exposure as we developed areas that our ancestors under-stood to be floodplains. For example, the Washington Monument in the US capital is located on a historic floodplain. From the Amazon, to the Mississippi, to the Yangtze rivers, we have seen images showing how levees, dams, and concrete channels intended to control water, failed by being overtopped by higher water levels. Even far inland, these super flood events are triggered by extreme rain and runoff to lower elevations. We've continued to allow vulnerable assets, homes, and farmland to be in harm's way for several reasons. If it has not flooded in a generation (about twenty-five years), there is a tendency to forget the catastrophe, or at least to dismiss it as a "freak event." Also, civil engineering projects such as the levees create a perception of lasting protection. Increasingly, it is clear that that sense of security is misplaced.[28]

Just one additional inch of water level can make the difference in whether a building floods or not. More catastrophically, there is a critical height where the failure of a levee or dam can bring deadly flooding to a community. It's the last inch (few centimeters) that matters. Inches can determine whether a sewer system or drainage network functions as intended. The extensive regional water-control system managed by the South Florida Water Management District is a vast network of canals, pumps, and sluice gates that "drain the swamp," provide drinking water, and are largely responsible for flood protection in the urbanized areas. One study showed that in South Florida, an 8-inch rise in sea level would require $1.25 billion just to install pumps to replace the gravity-fed water level control system that makes South Florida land, not swamp. It's just one example of the kind of unimaginable challenges, costs, and changes that rising waters will bring into play.

While on the topic, it's worth noting that pumps are not a panacea to solve rising waters. In addition to pumps being expensive and requiring maintenance, the energy required to pump water is not at all trivial. Anyone with a home swimming pool may be aware that it's likely the largest item on their electric bill. That's just to circulate water on level ground. Lifting water is substantially more work, requiring an enormous amount of power. During one meeting where pumps were proposed as the solution for Miami Beach in the face of rising sea level, a skeptical engineer showed me a back-of-the-envelope calculation that their entire electrical grid did not even have the capacity to keep the city dry by pumps. The same would apply most everywhere. Even if a larger electrical supply were installed, let's remember that generating electricity is one of the underlying demands that causes the greenhouse gas emissions that are warming the planet.

Another challenge to "standing our ground" is that in many coastal areas, the freshwater table will be compromised by saltwater intrusion.[29] In many areas, groundwater is the primary source of water for private wells and municipal utilities, replenished by rainwater, springs, and aquifers. In coastal areas, the rising level of saltwater is already starting to spoil the fresh water. Eventually this will require more desalination, a somewhat expensive process being used by an increasing number of communities. Desalination presents its own issues of power supply and what to do with the brine by-product.

Rising water levels will also impact landfills, hazardous waste sites, cemeteries, and all manner of sites that might not immediately come to mind. To reiterate, we need to consider not only the sustained flooding as sea level rises, but also the higher levels of short-term flooding that will occur during the routine king tides, storm surges, and deluge rainfall. Because they are on such different timescales, most people confuse them when thinking about the future, or do not see the need to plan for short-term flooding on top of higher base sea level.

Obviously, when sea level is several feet higher, many areas that are now dry land will be underwater permanently. Not only will their value plummet, but the ownership will also likely change. Generally speaking, one cannot own land past the ocean boundary in the same way as deeded real estate. It can vary by state, but the boundary for private coastal property is defined in terms of the high, mean, or low tide boundary, which will all change as ocean height rises.

Though little-noticed, the migration to higher ground has already begun, albeit quietly, because those abandoning vulnerable assets often don't make a point of why they're selling. Most coastal properties that have gone underwater, those that flood regularly, and those that are deemed to be soon at risk of flooding cannot recover. The decline in those real estate values will effectively be permanent. Given the interrelationship of property values and the financial markets, one has to ask how the write-down in property values might precipitate chaos in financial markets. In other words, would there be a cascading effect?

PROPERTY VALUES DISCOUNT FOR FUTURE FLOODING

This point about coastal property values never recovering as flooding worsens was clearly made in a 2017 report, "Life's a Beach." The title may seem playful, something that might appear in a tabloid, but it was published by Freddie Mac, the huge US government–sponsored Federal Home Loan Mortgage Corporation. To quote:

"One challenge for housing economists is predicting the time path of house prices in areas likely to be impacted by climate change," Freddie Mac Chief Economist Sean Becketti wrote. "Consider an expensive beachfront house that is highly likely to be submerged eventually, although 'eventually' is difficult to pin down and may be a long way off. Will the value of the house decline gradually as the expected life of the house becomes shorter? . . . Or, alternatively, will the value of the house—and all the houses around it—plunge the first time a lender refuses to make a mortgage on a nearby house or an insurer refuses to issue a homeowner's policy?" Becketti continued. "Or will the trigger be one or two homeowners who decide to sell defensively?" These threats are dangerous to homeowners because many times, a large share of their wealth is in the equity of their home, the report states. If these homes become uninsurable, the home values would sink, possibly even to nothing, and without hope of recovery.

Also, Freddie Mac warned if a home is literally underwater, it is not likely the homeowner would continue to make mortgage payments, causing a ripple effect in the housing industry and major losses for servicers and mortgage insurers, noting:

A large share of homeowners' wealth is locked up in their equity in their homes. If those homes become uninsurable and unmarketable, the values of the homes will plummet, perhaps to zero. Unlike the recent experience, homeowners will have no expectation that the values of their homes will ever recover.[30]

The key questions are how soon the write-down could occur and how precipitously. As noted above, there is an erroneous and widespread assumption that property values are only affected when sea level rises far enough to submerge the property. But as a series of articles are starting to note, coastal property will adjust downward preemptively. In fact, the process has started, as sampled by these articles:

- "Disaster on the Horizon: The Price Effect of Sea Level Rise," published in May 2018 in the *Journal of Financial Economics*. The peer-reviewed study concluded that vulnerable homes sold for 6.6 percent less than unexposed homes. The most vulnerable properties, which stand to be flooded after seas rise by just a foot, were selling at a 14.7 percent discount.[31]
- *Wall Street Journal* article, August 2018, stating, "Flooding has erased nearly $7 billion in value for homeowners in New York, New Jersey and Connecticut since 2005, a new study finds."[32]
- "The Risky Business Report" 2015. Between $66 billion and $160 billion worth of real estate is expected to be below sea level by 2050. By the end of the century, the range is $238 billion to $507 billion.[33]
- *The Economist*, August 17, 2019, in a double feature article, "A World without Beaches: One Way or Another the Deluge Is Coming" and "Climate Change is a Remorseless Threat to the World's Coasts," said, "The need to adapt to higher seas is now a fact of life" and "In thirty years the damage to coastal cities could reach one trillion dollars a year."[34]

Vulnerable coastal assets will likely not go down as rapidly as extreme financial market falls. In the case of accelerating SLR, I believe that coastal assets will go underwater in value sooner than most expect, but generally will not collapse. The more important point is that they will *never* recover. The desire by governments to compensate and indemnify property owners will likely prove to be an extreme challenge, given the scale and permanence of what is going underwater.

Our immediate instinct and effort is to "hold the line." Keeping the rising sea at bay forever will soon be understood to be prohibitively expensive and impractical. At present, the exodus of people from vulnerable properties and the direct economic impact from rising sea level is little more than a trickle. But those metaphorical "trickles" are already beginning to grow into streams and will eventually turn into rivers. The rate of increase will be partly driven by the accelerating melt rate in Greenland and Antarctica, of course, but the market's mood and increasingly widespread recognition of the reality will play an equally important role. In the future, the real estate axiom of "location" being the most important criterion for property value may become "elevation." This quote from the US Federal Reserve Bank of San Francisco closes the chapter succinctly, with authority:

> ... *environmental shifts such as rising sea levels and more severe storms, floods, droughts, and heat waves . . . will have increasingly important effects on the U.S. economy.*[35]

"Our old neighborhood sure has changed."

PART THREE

RISING TO THE CHALLENGE; LEADERSHIP AND VISION

9

SEEING THE "GLASS HALF FULL"— OPPORTUNITIES AHEAD

In the Chinese language, the word 'crisis' is composed of two characters, one representing danger and the other, opportunity.
—John F. Kennedy

Unstoppable, unpredictable rising sea level is really bad news. But as with many things in life, we have to work with what we have, or, as the saying goes, "play the cards we've been dealt." Our best option is to face this reality head-on. Rather than being depressed, we need to find ways to look at the situation as a challenging opportunity to motivate us to make the fundamental long-term changes that are required. In effect, it's a matter of seeing the metaphorical "glass half full" rather than half empty.

As rising sea level pushes the shoreline inland, the necessary adaptation will become the driving force for a large portion of the economy at all levels: locally, regionally, nationally, and globally. When policymakers and communities realize that SLR is accelerating and headed at least several feet higher, it will command our attention and become an inescapable priority, surely one of the world's biggest, resulting in all manner of construction and innovation. In fact, rising sea level *could* well be the greatest economic engine of this century—if we choose to approach it that way. Massive adaptation is inevitable. What we can control is what we choose to do about it.

KÜBLER-ROSS STAGES OF GRIEF

Knowledge of this new reality can send us to a dark place. Sustainability strategist Scott Nadler wrote an article in *Greenbiz*[36] that compares our angst about climate change with the Kübler-Ross stages of grief. From my experience presenting this case for almost a decade, I think Scott is spot-on. Realizing that we are about to lose favorite coastal areas evokes a real sense of loss that requires time to process. Having personally taught this material to many thousands of people in live meetings, I know that it has a literally stunning effect. Even the most analytical attorney, engineer, economist, or military leader can need time to process and adjust to the fact that we cannot stop the sea from rising. For those who have passionately championed the importance of environmental or "green" agendas as the solution, it can be especially difficult to accept the loss of vast coastal areas that we believed to be permanent.

The Five Stages of Grief, per Kübler-Ross:

1. DENIAL: Avoidance, confusion, shock, and fear
2. ANGER: Frustration, irritation, and anxiety
3. BARGAINING: Struggling to find meaning, reaching out to others
4. DEPRESSION: Overwhelmed, helplessness, hostility, flight
5. ACCEPTANCE: Exploring options, new plan in place, moving on

Short-term focus suits our brains' primal fight-or-flight system, which is wired to be alert to imminent danger. In an increasingly complicated world, daily headlines as well as personal and professional pressures reinforce this tendency to focus on near-term concerns, filtering out the less immediate. But we also have the capacity to take a longer view, such as with raising children, planning our careers, and looking ahead to retirement. The longer view of SLR must somehow rise to such a level of prominence. For that to happen, it helps if we are able to see how near-term sacrifices and thoughtful investments can lead to much better long-term results and even positive opportunities. Exploring examples of positive adaptation helps us realistically envision the future. We must move beyond cognitive dissonance (shock, disbelief, and avoidance), beyond despair and paralysis ("We're screwed!"), toward a perspective of practical planning. As with many major problems that we fear, some aspects of the future situation might be better than what we initially imagine.

TIME FOR TRIAGE AND FINDING THE POSITIVE

For our sense of purpose, stamina, and survival, we need to shift into triage mode. Anyone familiar with emergency response understands the three categories of where to focus efforts:

- Safe—not needing scarce resources
- At risk—might be saved with focused, urgent effort
- Unable to be saved—should not consume finite resources

In times of great social, political, or technological change, those who can see the larger trend are often best positioned to adapt, and even to thrive. This is true for individuals, as well as for companies, states, and nations. Looking back at past major transformations, those who overcame fear and anxiety and had the foresight to ride the waves of change to their own advantage became leaders. This happened during industrialization and during the advents of the automobile and highway systems, airline travel, modern communication systems, globalization, and most recently with computers and the digital revolution.

Fortunately, humans are amazingly adaptive. Individually, in small and large groups or teams, and particularly when it comes to our survival, we are incredibly inventive. There is an opportunity right now to be among the pioneers of positive adaptation in this new era, to find great purpose in embracing this new future with courage, and perhaps even with enthusiasm.

Once again, there is value to "beginning with the end in mind." When it comes to big, bold changes, such an approach helps us not to get bogged down in the constraints of politics, budgets, and the rut of doing "business as usual." Though those issues will have to be tackled, they are more likely to be resolved if there is a shared vision of the longer-term goal. We are going to have to build new cities with new infrastructures. There are communities in low-lying areas flooding today that urgently need to be completely rethought while there is time. Some iconic historical cities will have to be radically transformed, even if it threatens their historical purity. Over time, some will sink beneath the waves, a process that is already underway. We must raise street elevations and change building codes to achieve a sustainable positive return on investment, for public investment as well as private.

Such planning can include the "adaptive engineering" approach described earlier that allows for future modifications. For example, a structure might be built for 3 feet of rise but designed to easily adapt for double or triple that without having to be rebuilt from scratch. This is the value in looking further ahead and embracing the "begin with the end in mind" perspective. Investing in the future for ourselves and those that will follow always beats doom and gloom.

Anyone who expects me to describe exactly how to accomplish this will be disappointed. There is absolutely no way for any one person or any single book to cover this tremendous transformation to adapt to rising sea level. What will work in Portland, Oregon, will not work in Portland, Maine. What architects need to do is different than the tasks for accountants or attorneys. The only realistic goal of this book is to educate and stimulate the creativity, resources, and expertise of the diverse experts and members of coastal communities. In turn, you can help heighten awareness of the need for bold adaptation, rather than a shortsighted approach. What I *can* do in this limited space is give a few specific examples that show how bold approaches might add value and payoff in the future. The examples cover a wide range of situations, from mega-dams to tourism resorts, to stimulate thought for other applications.

DAM(N) GOOD IDEA?

Earlier, we looked at examples of visionary adaptations in Seattle, Galveston, and the Netherlands, reminding us that large-scale civil engineering was done successfully even a century ago. In Chapter Six, places were cited that might even be partially segregated from the rising ocean by dams, as a way to maintain the current shoreline. Let's drill down a little deeper to consider such a project. In 2016, I gave a talk in Copenhagen at a military conference. After explaining that rising sea level was inevitable—which, as usual, silenced the room—I drew their attention to the neighboring Baltic Sea, which touches nine nations: Denmark, Sweden, Finland, Russia, Estonia, Latvia, Lithuania, Poland, and Germany. I mentioned the wild idea of building the world's largest dam between Denmark and Sweden as a way to maintain the present sea level in the Baltic. Based on the map, it appeared to me that the best potential location was a place called the Kattegat, where the distance is relatively narrow. Though a barrier from Denmark to Sweden seemed like a crazy idea, they immediately realized the value of maintaining the current shoreline by having a dam to keep out the rising sea. The engineering challenge would be truly daunting, but the idea was also immediately thought-provoking and caused a lot of chatter.

As shown on the map below, the Baltic Sea has huge strategic importance for those nine nations. Any kind of a barrier to isolate the Baltic would deeply affect all those countries. A Baltic barrier would be by far the largest civil engineering project in the world. Theoretically, it would maintain the current sea level and shoreline along the Baltic coast far into the future while maintaining ship traffic via a system of locks. The only other option is a traditional dike or levee around the entire Baltic, which would have its own challenges and cost, given that the shoreline of the Baltic is roughly 8,000 kilometers long (about 5,000 miles). Consider the truly enormous value of property and infrastructure around this great sea that would be lost with even 3 meters (10 feet) of SLR, which could move shorelines far inland, and it becomes immediately clear that it's a time for big ideas such as these.

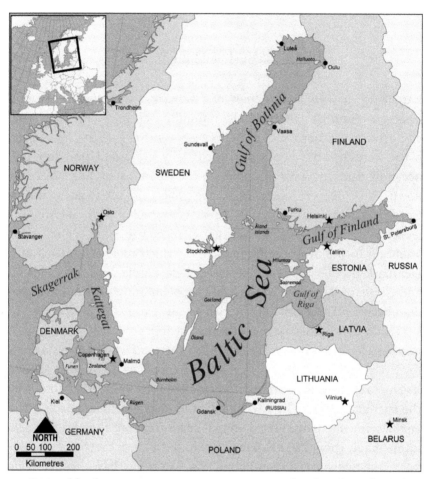

It would take many years to get acceptance for the idea of a Baltic Barrier. Just the feasibility studies would take years. Then it might take decades for engineering and environmental studies and approvals, and then construction after that.

Since SLR is unpredictable and could increase rather rapidly as early as midcentury, there is merit in discussing such concepts now, long before they will be needed. Each of the bays, gulfs, and seas cited in Chapter Six, such as San Francisco Bay or the Mediterranean, would have different challenges. Without doing a thorough evaluation, based only on the width of the openings that might be dammed, the easiest barrier to construct would be across the very narrow Dardanelles between the Aegean and the Sea of Marmara on the route to Istanbul; the most challenging site would likely be the Gulf of Mexico. The point is that it's time to start thinking outside the box now, decades ahead of the need.

Just as this book goes to publication, a dam concept was publicly proposed, at a larger scale even than my suggested Baltic Barrier. NEED—the Northern European Enclosure Dam—is a provocative and ambitious proposal to build two enormous dams even farther out into the Atlantic Ocean, protecting even more of Scandinavia, Europe, and parts of the United Kingdom. The proposal by a Dutch government scientist was published in January 2020 in the prestigious *Bulletin of the American Meteorological Society* and quickly made its way into *Dezeen*, an architectural journal,[37] with the headline, *"Scientist Suggests Damming the North Sea to Protect Europe from Climate Change."*

Concepts for such mega-dams will no doubt be explored more and more seriously as the unstoppable rise becomes clearer. It remains to be seen if they are realistic. When we turn our attention to creative ways that we can design for a challenge, a process that architects call a *charrette*, something quite interesting happens. What had been a daunting dilemma turns into an energizing challenge with some competitiveness. I have seen this happen time and again in the presentations I give to engineers, architects, contractors, investment bankers, municipal planners, and attorneys. Most of these professionals had not been exposed to the facts about rising sea level or had been operating under the erroneous belief that SLR could be soon halted. Even those who knew the facts may not have been mentally ready to think about how to address the challenge, a form of the aforementioned cognitive dissonance.

At the end of August 2017, I gave the keynote talk to an audience of more than 400 military engineers in Jacksonville, Florida. Many were from the US Army Corps of Engineers or had similar positions in the other service branches. When I explained that sea level rise was unstoppable and that we could see as much as 10 feet of SLR within the next hundred years, the room became absolutely quiet. Then I mentioned that the very narrow entrance to the St. Johns River right there in Jacksonville might be one of the easiest places in the world to erect a storm surge barrier that could protect the huge low-lying delta region of greater Jacksonville. I pointed out that such a barrier would be particularly effective to protect them from any hurricanes from the east, which can push water into the river mouth and flood the huge Jacksonville area. For engineers, this was a problem that they could tackle. Immediately, they became fully engaged in conversation; some even began to sketch and compare ideas.

> Eerily, just eleven days after I described that scenario to that engineering conference in Jacksonville, Hurricane Irma followed that script exactly. The extreme storm surge came right up the St. Johns River, badly flooding most of greater Jacksonville for weeks, including the navy base and the hotel and conference center where I had spoken. Sometime later, the senior engineer from Naval Station Mayport told me that my hypothetical flood scenario came to mind frequently while he was dealing with the real thing. That's now the third time I have publicly described a severe theoretical flood scenario that actually happened within weeks, straining explanations of coincidence. While curious, it's not any kind of supernatural or prophetic power (though I have to admit, I now hesitate when describing futuristic scenarios). I believe the synchronicity (coincidence) comes from my looking at the big picture of flooding with rising seas and changing weather patterns, without being constrained by what we have experienced.

Several major global engineering firms have developed the capability for large-scale engineering solutions to accommodate rising water levels. In my discussions with them, it has become clear that with few exceptions, most clients only want a limited solution, perhaps looking out twenty years. Some cities are looking ahead to the second half of this century and a possible rise of 2 to 3 feet. Examples include New York, Boston, Miami, San Francisco, Rotterdam, London, and Singapore. To be clear, they do not have the problem solved, nor are they even settled on the solution, but at least they are working on it. That's commendable and a good start. Yet if all coastal communities worldwide are to adapt, we need to think bigger, much bigger.

For our purpose here—getting our minds around the process and the hurdles to moving to higher ground—it is useful to think conceptually and at a larger scale. For example: Is it realistic to think that people will only move for survival, or through some government or facilitated program? Or will they migrate and disperse gradually over a period of many years, as they do now after major hurricanes and floods? Affected municipal governments will struggle, doing their best to stay solvent and deal with the needs of people and businesses to maintain the tax base. But in the face of rising sea level, most residents of low-lying coastal communities will adapt—most likely moving away—sooner or later in their own best interest.

But there is another approach that is largely missing so far. Imagine a city where thoughtful leaders see the handwriting on the wall and develop a citywide vision for a future when sea level is 5 feet higher *or more*. To get people to think creatively, unrestricted by their current structures and infrastructure, they use the far-off year of 2100, far outside people's immediate concerns. A futuristic view of this iconic city might include elevated structures, stilt homes, neighborhoods on platforms, or floating structures of some design that protects them from ocean waves. Even conceptually, such a bold community redesign would involve diverse experts and resources, including architects, engineers, landscape architects, economists, planners, and artists. The goal of such a futuristic vision would be to envision a thriving community and excite and inspire people to think boldly. I believe that such an exciting vision could be inspiring enough to encourage us to do the hard work today in order to keep the economy and infrastructure functioning for years as we navigate the stormy waters of the future.

In this view, I believe the key is to stay focused on developing a long-term vision and to do our utmost not to get derailed by short-term politics, personal interests, and cost concerns that inevitably come into play as soon as anyone anywhere talks about major infrastructure change. The practical, the short-term, and the medium-term challenges will all have to be addressed, but should not thwart the process of at least *thinking big*. In part, we face an engineering and economic challenge, but this is really a large-scale planning and design challenge. While many will roll their eyes and say that such adaptation is impossible, it is not only possible, it is inevitable. As the ice sheets melt and the sea rises, we will be forced to get over our attachment to the current shoreline and do what humans do best: adapt and exercise ingenuity, particularly when facing the need for survival and security for ourselves and our loved ones.

Consider what will happen when one city bravely stakes out the position of being the first to design for 2 or even 3 meters of SLR. (Hopefully my use of the alternate metric system throughout this book has my fellow Americans realizing by now that 2 to 3 meters is 7 to 10 feet.) The message and impact that one city has boldly started to plan for as much as 3 meters will be international news. It would serve as a model and inspiration for others. No doubt the naysayers will give lots of reasons this cannot work. In terms of vision and boldness, it would be the equivalent of President John F. Kennedy setting the goal of landing on the moon by the end of the 1960s. It's easy to forget that that audacious plan was controversial. It was thought to be irresponsibly expensive and risky. The results of the ambitious space program are now well known and generally considered worth the cost.

Returning to the concept of a future coastal city that can thrive at much higher sea level: Whichever city or region starts such a plan for multi-meter SLR will get to proclaim itself a city of the future, one that will not only put a floor under its property values but will likely have premium value simply based on the premise that they *have a plan* for the future. It has been demonstrated countless times, all over the world, that investment values and real estate prices respond to bold vision.

FROM WRITE-OFFS TO VALUABLE NEW ASSETS

Take a simple example of the economic concept applied to one hypothetical project in a coastal area where there is increased flooding following the more frequent king tides due to rising sea level. Assume it's a twenty-year-old beachfront resort that is worth thirty million dollars, in addition to the value of the land. Assuming the standard practice to depreciate buildings over thirty years, the resort might now be worth ten million dollars on its balance sheet financial statement. That amount would be written off as an expense as that asset floods more. The company might build another resort in another place on higher ground using some creative design and landscape architecture to make it much less vulnerable to future sea level rise. Perhaps the new resort costs forty million dollars. The companies designing and building it will have forty million dollars of new business. The resort company would have a much more durable asset on its balance sheet, perhaps with lower costs for maintenance and insurance, yielding good profit. In this hypothetical example, ten million dollars of invest-ment would have been written off, perhaps with some benefit such as a credit against taxes. But the replacement property creates forty million dollars of new economic activity for design and construction and results in a higher-quality long-term investment. You may quibble with certain details of my scenario, but the point is that rebuilding or relocating specific properties and communities to higher ground can be a huge net positive in financial terms.

Now imagine scaling that single coastal resort by hundreds of thousands of coastal assets, or worldwide, millions. Along with adap-tation, new techniques and technologies will be developed. Fortunes will be made, likely offsetting the losses, at least at the macroeconomic level. As with any disaster, it is our optimism, creativity, resiliency, and survival instinct that propel us forward. The sooner we begin planning for the inevitable adaptation, the better the outcome for the largest number of people.

For every billion dollars literally put underwater by the rising ocean, we will create replacement assets of comparable or greater value. Following the above scenario, I think the case can be made that it will be an economic positive.

WE MUST RISE WITH THE TIDE

"A rising tide lifts all boats" is an aphorism often stated by President John F. Kennedy. For him it was a message about the economy, but it is particularly apt here. Rising sea level will be the biggest economic driver this century. In many ways it will displace a lot of our pres-ent-day priorities. This ever-rising tide will not bow to legislative edicts, politics, or financial constraints. Without question, it presents a huge problem. But it is *both* a crisis and an opportunity. If we act sooner and smarter, there will be more opportunity and less crisis.

10

WHO WILL LEAD?
POLITICIANS,
PROFESSIONALS, OR PUBLIC?

*The problems of the world cannot possibly be
solved by skeptics or cynics whose horizons are
limited by the obvious realities. We need men
who can dream of things that never were. Change
is the law of life. And those who look only to the
past or present are certain to miss the future.*
—John F. Kennedy

"Wicked problem" is a term used by professional planners for those challenges that are difficult or impossible to solve. "Wicked" means more than very difficult. With a wicked problem, both the goal and the solutions defy clear definition, in part because the "goalposts" will move depending on the actions we take. Rising sea level is a very wicked problem.

In the face of unstoppable, unpredictable, accelerating sea level rise, the key question is: What can we do, both as individuals and collectively? Though the general belief seems to be that the key decisions need to be made by big government, I have a somewhat different view.

While there are clearly things that local, state, and national govern-ments should do, individuals also have an important role to play. For those ready to be in the vanguard of early adapters for SLR or those who are ready to step in as visionary leaders, it will help to look at poten-tial efforts and policies, from several different perspectives: personal, business/professional, community/local government, and higher levels of governance.

Moving to higher ground—both literally and metaphorically—is serious stuff, requiring more than feel-good measures. If we want to reduce the impact of SLR, we need to be honest and realistic with our-selves and others. The first step is to identify which efforts will lead directly to solutions. At a macro level, I find it useful to distinguish four separate categories of climate and environmental effort. Though there is overlap, they are quite distinct:

1. **Reduce CO_2 emissions to slow the warming, the melt-ing ice, and rising sea**, yet realizing that personal and local efforts to reduce CO_2 emissions only have an impact as part of a global total. Even with great effort, there will still be a very long lag time before it affects sea level.

2. **Prepare for the more frequent flood events** that are already occurring by being more resilient. We accomplish this by shoring up our ability to resist and endure when the water rises rapidly and to recover when the water goes back down.

3. **Prepare for long-term sea level rise** by changing build-ing and zoning codes, recognizing the durability of build-ings and infrastructure. Plan for rapid SLR acceleration and the likely abrupt increases in the coming decades. (This is of course the primary focus of this book.)

4. **Address the multitude of other critical environmental issues**, including clean air, safe drinking water, recy-cling, ending the scourge of plastics in the ocean, coral reef protection, wildlife and ecosystem conservation and restoration, etc.

If your greater concern and higher passion is for categories one, two, or four, hopefully you can see the way that the rising sea level issue complements and supports your priority.

> "Saving the planet" is a common message these days, and for good reason, but it can be difficult to separate the legitimate from the spin and hype. For example, most hotels promote reusing towels; many suggest that this combats climate change, with some even hinting that less laundry can save the polar bear's habitat. That's really stretching the truth. Most savvy travelers realize the motivation is more to save on hotels' significant laundry expenses. Although reducing energy consumption and the environmental impact of the phosphates and nitrates commonly found in laundry detergents is a good thing to do, I believe we need to be honest and stop spreading myths and misunderstandings about such important issues.

SLR: THE GLOBAL ISSUE THAT ONLY HITS LOCALLY

Rising sea level may be a global issue for some, but for most people, it will be a local issue, rooted in whatever place they call home. Our homes are often our biggest assets, as well as being essential to our sense of community and security. We are deeply attached to places, those where we grew up, where we raised children, where we have memories, and perhaps where parents or grandparents are buried. Psychologists describe this human characteristic as "place attachment." Place attachment is an important part of why we do not want to retreat and move to higher ground. The physical world keeps us grounded in more ways than one.

Furthermore, retreat is associated with defeat, as in a military battle. "Standing our ground" is not only the "manly thing to do"; we associate it with defending our home against intruders and protecting our assets, a positive thing. But a rising ocean is different, an enemy we have never seen before.

INTELLIGENT ADAPTATION

In my first book on this subject, *High Tide on Main Street,* I identified five points for *intelligent adaptation* as a macro perspective, which many readers found helpful. The underlying facts have all been covered in this book, but it seems appropriate to include the numbered list here:

1. **Act with a long-term perspective.** Plan ahead so that investments have adequate scale and are more durable.
2. **Accept that there is a range of projections** for sea level rise this century. Do not let yourself get paralyzed and confused by the range of estimates nor wait for there to be settled precise amounts.
3. **Consider geology as well as topography.** Places with porous limestone or subsidence have different problems that need to be considered for future planning and investment.
4. **Recognize the finite future of government bailouts.** Despite best intentions, expecting the government to absorb the rapidly growing risk of future sea level rise and associated risk for the long-term is unrealistic and will lead to increased exposure.
5. **Anticipate property devaluation.** Assets that are vulnerable to SLR could be devalued far ahead of the actual flooding.

ASSESS YOUR VULNERABILITY: A CHECKLIST

Doing a vulnerability assessment for rising sea level is highly recommended. That could take the form of a self-assessment with a series of simple questions or a comprehensive professional engineering analysis. Your choice likely depends on your assets at risk, your proximity to the rising sea, and other flood factors. For those inclined to do it on their own, I have created a basic checklist that may be helpful. "Keeping Your Assets Above Water Checklist" is in the back of this book and can be downloaded for printing as a worksheet. Considerations include such obvious issues as the elevation of your lowest window or door opening; the risk of flooding from uphill sources; exposure with access roads, electric, water, and wastewater utilities and vulnerabilities of lakefront and tidal river locations and waterways. Your age, net worth, and risk tolerance are also important factors.

IT STARTS WITH EDUCATION: WE ALL NEED TO BE TEACHERS AND STUDENTS

Education may sound like a soft tactic, but this is such a revolutionary and misunderstood problem that education is the critical first step. Even professionals are frequently confused, for example, thinking that reducing emissions can stop rising sea level. Anyone can be part of the process to improve our understanding and make a difference. If the engineers, architects, lawyers, and financial community do not understand the issues of long-term SLR, we will not develop good adaptations. We simply cannot afford to waste money and time on misguided, shortsighted efforts. At this moment you are educating yourself. By sharing this new information with others, you can start to exert influence from the bottom up, encouraging professionals and local governments to listen.

Rising sea level is extraordinary in that every person on the planet has things to learn on the topic and things they can teach others. Having personally given hundreds of talks, including to some supersmart people, I can't recall any where I was not able to open their eyes to some fact or implication of ever-rising seas. Equally, I continue to learn about the facts, impacts, and sociopolitical aspects of this phenomenon. That's the reason I dedicated this book to those teaching the topic and those who will join us as teachers. In this great adaptation, everyone can be both student and teacher. It is incumbent on us for our own benefit, and more importantly, for those who will follow us.

> To help you share this information with others, there is a link to download a free slide set in PowerPoint or PDF format that you are welcome to use under "Other Resources" at the back of this book.

THE "RETREATERS"

A few years ago, I happened to meet Bob Gremillion, who had recently retired after two decades as publisher and group VP at Tribune Publishing, owner of the *Sun Sentinel*, the major local newspaper in Fort Lauderdale. "Lauderdale" has so many canals and waterfront homes that it's described as the yachting capital of the world and the Venice of America. His newspaper had given decent coverage to the growing concern about rising sea level and "sunny day flooding" from the king tides. After hearing me explain rising sea level for a television interview, he introduced himself, saying he had not understood that sea level rise is unstoppable and would probably accelerate. He got a copy of *High Tide on Main Street*. Over the following weeks, we had several calls and soon met at his canal-front home, positioned a rather safe 13 feet above sea level. Perhaps because he was originally from New Orleans, this issue really took hold with Bob. He expanded my audience, recommending me as a keynote speaker, describing me as "the Paul Revere" about the rising sea. Before long, Bob and his wife, Sally, put their wonderful waterfront house up for sale. Now they live in the North Carolina mountains, two thousand feet above sea level. They are among the growing number of people I know who have literally moved to higher ground, some of whom glibly refer to themselves as "retreaters." Although a few are concerned about actual flooding, most are concerned about the general devaluation of low-lying coastal property that could affect them long before the water arrives.

As people become more aware of the risk of flooding, they will reduce their risk by protecting their assets. Even without full retreat, that could mean something as simple as sealing a low window opening. Or it may require elevating the entire building. As noted earlier, however, it's not just about protecting your building; the neighborhood, the utilities, the infrastructure, and the regional risk of flooding may turn your property into what is termed a *stranded* asset—something with intrinsic value but that you can no longer fully access. Each person needs to evaluate their level of vulnerability, their community's exposure, and their own risk tolerance in order to make an informed decision.

Many in the process of retreating do not voice their motivation for moving, for fear of triggering a decline in local property prices while they are selling. I made reference to this phenomenon in Chapter Eight. While there is a tendency to avoid looking out too far into the future, markets do take future possibilities and risk into account. This happens with stocks, bonds, and all manner of investments. If we think the future for a given asset is good, the price will tend to increase; the opposite is true as well. Recall that I quoted the Freddie Mac economist who proposed several different possible triggers for a collapse of coastal asset values. With rising sea level, there is an added challenge that the problem seems so distant and most people do not like negativity, referring to it as "doom and gloom." The point is that it may behoove you to do your own evaluation early, rather than taking a wait-and-see approach, by which time it may be too late to get the full value out of your home.

DIFFERENT PERSPECTIVES

It's important to recognize that politicians and property owners are not entirely aligned. While both groups are dedicated to their communities, elected leaders are responsible for the well-being of a community as a whole. Individuals, however, must determine what is in the interest of themselves or their families, which may involve leaving a place they love if the value of their biggest asset is at risk. In other words, even with intelligence and sensitivity, the sociopolitical realities of a community will put different people in different roles. Battles over zoning and public policy have often been contentious. Rising sea level raises the stakes even higher.

"SALT" AS SMARTER PUBLIC POLICY

We need to develop and advocate for programs that encourage smart policy, including the managed or personally directed retreat that will be inevitable in most coastal zones over time. Just as an example, I developed a conceptual program, Shoreline Adaptation Land Trusts (SALT), that could be a useful vehicle to encourage the gradual voluntary shift inland from the coast. Following established principles of donating land to conservation land trusts, it outlines parameters for owners of vulnerable properties to donate them to a public trust, allowing them to enjoy the use of the property for their lifetime and getting tax benefits for the donation. The three-page description, created for *The Institute on Science for Global Policy*, is available online.[38]

In Chapter Six we briefly looked at adaptations in the face of unpredictable rising seas, including mobile homes, houseboats, amphibious homes, and all manner of ways to elevate structures on pilings and jack-up legs. Living on a boat is another option. The maritime life is not for everyone, but it's easy to foresee growth markets in all things that float, from true ships to anything on pontoons. Along with boats and floats, there will be growth in all manner of shoreside support, from marinas, to lagoons for protection from waves, to dry dock and repair facilities. I have even heard of plans for floating airports. That said, we must not delude ourselves into thinking that a majority of the tens of millions of people now living right at sea level are going to move out to sea. With troublesome obstacles, including exposure, coastal storms, cost of construction, and maintenance in the marine environment, I do not see it as a realistically scalable option for the tens of millions exposed to rising seas.

PROFESSIONALS—CHANGE THE BUILDING CODES

Building awareness and support among professionals and the business community is another avenue for getting the political leadership and the public to embrace the huge change that is necessary. Architects, urban planners, engineers, attorneys, insurers, the financial community, and the military potentially all have a great deal of influence on this issue. Some of those professionals have shown interest in learning about adaptation as a professional challenge and positive opportunity. It is important that they see themselves as positive change agents. In their daily work, they have the opportunity to incorporate awareness of rising sea level, looking for ways to adapt.

Perhaps an even stronger motivator for those same professionals is the risk of negligence and liability exposure. Just as one hypothetical example, consider all the parties involved in a large commercial building in Miami with an underground parking garage that will surely flood in the future. It is not hard to imagine career-destroying lawsuits naming all parties involved—the architects, engineers, contractors, lawyers, and economists who are party to such a project. The same could be said of large bond issues, residential real estate developments, a new convention center, multibillion-dollar public-private partnerships that now finance expensive infrastructure—the list is long. A wide array of professionals can lead the way with adaptation as the safe precautionary path.

In addition to the professionals, their academic counterparts will need to develop new content for coursework and degrees. Young innovators will get to make their mark on various professions, with new approaches that recognize the rising tide as a "game changer."

Accounting Rules. To show how far-reaching the effects will be on professions, beyond the obvious involvement of economists and the financial sectors, the accounting profession will need to change policies. At present, coastal land stays on the balance sheet, unchanged, usually at purchase price. This was reasonable on the basis that land was permanent.

On the other hand, buildings are depreciated or amortized as a way for companies to take a portion of the asset as an expense and recognizing a long-term decline in value. With the realization that over the next century all coastal land will need to be written off and expensed in some manner, unstoppable rising sea level will cause parts of the Generally Accepted Accounting Principles (GAAP) to be rewritten.

Architects are particularly well positioned to lead the charge. Increasingly they are designing for the possibility of flooding. Generally, the issue of flood height is determined by the building code, specifically the height requirement for the *finished floor elevation* (FFE), but in some cases, the clients are requesting it. Miami architect Reinaldo Borges is among those pushing hardest for adaptation to anticipate rising sea level. With passion and perseverance, he has made some inroads, but readily admits that people "build to code," referring to the building codes that exist at state, county, or municipal levels. "To get change, it has to be required in the regulations," he told me. He is right, of course. Although there has been some increase in FFE, he is pushing for more. He has also managed to create some innovative designs that go beyond code requirements. In addition to the obvious raising of the ground with fill and use of columns to elevate structures, his innovations include putting vulnerable equipment on higher floors and allocating lower floors for parking, which can be allowed to flood during short-duration extreme events.

JUST ASKING THE RIGHT QUESTION CAN ADVANCE ADAPTATION

It can be frustrating to try to get companies, governments, and even professional organizations to change quickly. But having worked on this for quite a few years, I have found a few techniques:

- Speak to them in their culture and idiom
- Identify aspects of self-interest

- Raise the liability threat of negligence and malpractice (covered just above)

With SLR being so huge, vague, unprecedented, and slow-moving, it's important to find a point of connection or an opening that engages people. Consider the following example. I was asked to brief all the flag officers (admirals) of the US Coast Guard on the issue. Ahead of my taking the podium, one of them politely told me that he "was skeptical of this climate change stuff." During my presentation, I kept glancing at him. His reaction was quite visible, going from arms crossed and leaning back in his chair, to leaning forward, eyes widening, and nodding. I made the point to the assembled officers that in a typical twenty- to thirty-year career, there would be changes that are almost unimaginable today. Several commented that they were already seeing higher water levels at their shore facilities. I always try to introduce a little humor on such a somber topic. For them I suggested that if their name and mission is to guard the coast, and their motto is *semper paratus*, meaning to be always ready, they better know where the coast is going to be. It got a good little laugh.

But what really made an impression was something I threw in as an example, *ad lib*. I brought up the subject of bridge clearance heights, something I touched upon a few chapters back. The Coast Guard has to approve those for all navigable waterways so that cargo ships and sailboat masts can pass beneath the bridge. I suggested that it would make sense to take rising sea level into consideration when looking at vessel clearance criteria. Bridges nominally have a design life of a hundred years; many last much longer. If a particular new bridge called for a clearance height of 200 feet, I suggested they should require an additional 10 or 20 feet to ensure its functionality over its useful life. With only a minor impact to the engineering or construction cost, that modest change could make a huge difference in the latter part of this century as water levels rise. In addition to being a practical issue, it was easy to visualize, and something they could implement quite quickly. Most importantly, it showed the value of good planning and "looking over the horizon." That caused smiles and relieved the discomfort of an insolvable problem. The same concept applies to all coastal infrastructure design and engineering. It was gratifying that even that admittedly skeptical admiral approached me following my talk and gave me a thumbs-up.

For the commandant, Admiral Paul Zukunft, it led to a mission to Greenland to see firsthand what is happening with the glaciers and to meet with his Arctic counterparts to consider the implications of unstoppable rising seas and changing coastlines. Again, my point is that different people or organizations will be engaged by different elements of this challenge, which is beyond our historical frame of reference. For something as daunting and foreign as rising sea level, it is good to find the "hot buttons" and points of connection for each audience.

FOR CONSULTANTS—A MAGICAL QUESTION

Ecology and Environment Inc. (E&E), an environmental engineering company now part of WSP, provides another good example of how to integrate sea level rise into planning. The firm was working as a subcontractor on an expansion project for a secure military facility on a remote island. At the meeting with the contractor and the military contracting officer, the E&E lead asked how much sea level rise they were planning for. The reaction indicated there was none, but they asked what he recommended. He quoted my guidance to "plan for the first 3 feet of sea level rise as soon as possible," given the uncertainty about the actual rate. The military contracting officer asked the contractor how much that would add to the cost of the project. With little hesitation, the contractor said it just meant adding another four rows of concrete block to the structure. Given the overall cost of the project, it would be so small as to be insignificant. With that, the military lead authorized it. Amazingly, it was just a matter of posing the right question. The question "How much sea level rise do you want to plan for?" turns out to be magically effective.

Planning for long-term SLR will rarely be on a project team's list of priorities. But when they are specifically asked how much they want to plan for, the "safe position" becomes one of planning for greater resiliency and durability. The difference is subtle but huge. The trick is to ask how much SLR clients are designing for, rather than to start by telling them what they should do. Where the client seems resistant, I recommend writing on the specifications or drawings, "plan for no sea level rise" and getting them to initial it. Not many people want to "own" that position; that forces them to address it head-on.

Professional planners are comfortable with thirty years or longer as a good "horizon" for infrastructure. Though that would not account for the full anticipated amount of SLR, it would certainly inform their thinking as they look ahead. But in most cities, the planning department is one of the first for staff cutbacks when budgets get tight. Pressure and priorities usually force them to focus on near-term issues—growth, new neighborhoods, traffic planning, utilities, school access, and so on—not decades in the future. With the exception of a few cities like New York, Boston, Miami, and San Francisco in the US, little attention is paid to planning for rising seas. Even in those places, there is little support for looking beyond a few feet of SLR.

South Florida has demonstrated one notable example of leadership regarding this issue. For more than a decade, the four southeastern counties of Palm Beach, Broward, Miami-Dade, and Monroe (the Florida Keys) have collaborated to assess and share their findings on projected sea level rise. Pushed into existence by one strong leader, the late Kristin Jacobs, the Regional Climate Change Compact annually assesses the science so that the region has a unified guideline against which to plan, to avoid inconsistencies among adjacent jurisdictions. As previously shown in the online Deeper Dive Note #4, in 2020 the collaborative committee adopted a rather aggressive recommendation that estimated 1 to 2 feet of rise by midcentury and up to 7 or 8 feet by the end of the century. They are to be commended for their efforts to address the issue. The problem remains how to actually plan for adaptation in accordance with that guidance.

Instinctively, people want to know what "the government" will do. However, even in cities where there is a lot of talk about dealing with rising seas and more flooding, most governments will not raise the elevation requirements for buildings, roads, and infrastructure, or may timidly raise them a small amount. Many elected officials with limited terms in office have little appetite for painful choices that won't yield visible results right away.

MOVING CITIES—REALLY?

From time to time, I hear references to "moving a city." This is unlikely to be a widespread solution. First of all, there are very few situations where it is possible to move an entire city. In *High Tide on Main Street*, I cited one example, where the small Central American country of Belize moved its capital inland to a new city of Belmopan, at an elevation of 250 feet. It was a smart, carefully planned response to a devastating hurricane in 1962 that destroyed their coastal capital. But the population is less than twenty thousand. Even in that case, most people commuted to work in the new city, but kept their residence in the old capital. That's an exceptional example. Rarely is there the land or the will to make such a move.

In Louisiana, the community of Isle de Jean Charles is steadily sinking into the bayou due to land subsidence and rising sea level. Even with some forty-eight million dollars of funding to relocate the town of a few hundred people, a large percentage do not want to leave. That example underscores our great resistance to move, as well as the incredible cost when the government takes on the responsibility. The legal precedent of such a move has not been adequately considered. Or even the question of who is responsible for assessing our risk and vulnerability. It's an immensely complex issue even in the smallest of communities. With larger communities, the difficulties are infinitely greater. Where could the populations of many millions in Miami, Fort Lauderdale, New Orleans, Jacksonville, Shanghai, or Calcutta/Kolkata *go*?

LARGE-SCALE EXAMPLE NEEDED

Planning our future path will take boldness, large enough scale, a "one-hundred-year vision," and a willingness to do it without consensus. My belief is that revolutionary projects require a few special things:

- **A well-defined goal**, with parameters
- **Leadership**—a strong CEO at the helm
- **A selected, dedicated team** to work towards the goal— as opposed to a normal community that inevitably has

diverse points of view and will usually spend a lot of time
in discussion and even argument seeking consensus
+ **Adequate resources**, generally meaning capital
+ **Sufficient scale** to implement the concept; size matters
+ **Freedom to act**—insulation from the inevitable critics,
threatened interests, and the political fray

Consider how some of the world's great revolutionary undertak-
ings met those criteria in some fashion: Putting a man on the moon
and safely returning him to Earth; building the atomic bomb; creating
the United States of America as a new form of governance and society; the modernization of China. Of course, there are many others.
I believe that successful execution of bold mega-projects requires all
those characteristics.

The United Nations? Some point to the UN as a possible
leader. Certainly the challenge of rising sea level will impact
their missions and programs. Their continuing focus on
climate change is quite commendable but cannot yet be
deemed successful. I cannot see how they are the vehicle to
lead on rising sea level, in terms of global adaptation. For
example, the UN's 17 Sustainable Development Goals are
commendable but do not even recognize the need to adapt
to rising sea level.

LEADERSHIP: QUEEN BEATRIX AND PRIME MINISTER LEE

To really change our understanding of how to plan for future SLR on
the order of 3 meters (10 feet) or more, I think it will take a bold example, at least at the city level. Former Queen Beatrix of the Netherlands
provides one example of the kind of real leadership that is needed.
Obviously, flooding and coastal defense are top of mind for the entire
Kingdom of the Netherlands. The country's very name means low-
lands, with almost twenty percent below sea level and half the country

less than a meter above sea level. Over the last two decades, the warming temperatures, melting ice, and accelerating sea level rise have been of increasing concern there. In 2008, Queen Beatrix called a meeting with climate scientists Drs. Paul Crutzen and Robert Corell, along with Dr. Karin Lochte, the director of the Alfred Wegener Institute, to discuss the question of climate change and particularly what rising sea level means for the Netherlands. It was quite unusual in that it came from the monarch and not the government, headed by the prime minister. Sessions with the relevant Dutch government ministers ensued over the next year and a half. When it was eventually established that it was not possible to specifically define the upper limit of global sea level rise (SLR) this century, the queen decided that the responsible thing to do was to plan on a precautionary guideline of 3 meters of SLR (10 feet) over the course of this century. The country was to commit on the order of 1.5 billion Euros annually (US $1.7 billion) to improve resiliency and to adapt.[39] That figure represents almost four-tenths of one percent of the Dutch national budget. Back then, more than a decade ago, some of the hundreds of engineers, scientists, and people in government agencies felt she was going too far. Even today, many still do not want to contemplate such a future and appear to have done what they can to dilute her planning guidelines. It's my perception that, generally speaking, governments and large engineering companies are uncomfortable with sticking their necks out. Her decision to convene a few select experts, give direction to governmental agencies, and ultimately make a bold visionary decision showed real leadership. (Queen Beatrix abdicated her throne in 2013 to her son, King Willem-Alexander, who appears to be aligned with her thinking.)

Boston is a good American example of a city with high vulnerability and complexity due to the shape of its harbor, many low-lying areas throughout the city, the famous Charles River, the Back Bay, and so on. Even Cape Cod works as a long arm to funnel water and amplify water heights into the historic city under certain conditions of weather, tides, and currents. In Chapter Six I covered some of the recent efforts to look at the city's risk of flooding futuristically. It's too early to know for sure, but Boston's process shows promise of planning and implementing for longer-term SLR adaptation. It's certainly further along in its thinking than most other cities in a similarly vulnerable position, though New York City also has done good work in this regard.

Closing this chapter focusing on leadership about SLR, I have to comment on the United States, which has generally not made it a national issue. As a candidate and as president, Barack Obama did make "slowing the rising oceans" a goal, but was unable to deliver. The clearest and most consistent leader at the federal level has been Senator Sheldon Whitehouse from Rhode Island, "the Ocean State." Rarely a day goes by in Congress without him warning about the worsening threat of climate change, and rising sea level in particular. Rhode Island, by far the smallest state in the US, is extremely vulnerable to rising seas. SLR is a real priority for his constituents. Senator Whitehouse bravely fights to make it a national issue too.

On the other side of the world, Singapore recently staked out a very bold position. Since Singapore became independent in 1965, it has increased its small-land area by twenty-three percent using techniques similar to those of the Dutch. This city-state is a regional powerhouse despite having a population of less than six million. On the world stage they have earned high respect for financial soundness, a strong social safety net, and a harmonious multiethnic culture. They have made the problem of rising sea level their highest priority. In August 2019, Prime Minister Lee Hsien Loong said the country would prepare for up to 4 meters (~13 feet) of SLR. The initial estimated cost is more than 100 billion Singapore dollars (approximately US $75 billion). That is the clearest and strongest possible policy statement to date regarding SLR of any nation.

HOW TO FLOAT THE FUNDING

Envisioning the future this way is essential to creating communities of tomorrow. It will take leadership and vision. An immediate question that arises is how to pay for it. Perhaps there will be new superbonds. Some have suggested a special financial investment, catastrophe "cat" risk bonds, as an example. However, those are generally looking at extreme events like hurricanes. Recall the difference between a localized, random event, for which the risk can be spread among many participants in the insurance pool, versus SLR, which is slow, universal, and permanent. There are now "green bonds" dealing with the financing of renewable energy, freshwater resources, and making cities more resilient to flooding events like hurricanes and inland deluge; perhaps financing the adaptation to SLR might build upon them.

Adapting or relocating all coastal communities to higher ground is going to be—with little hyperbole—astronomically expensive. Some have pointed to the controversial Modern Monetary Theory (MMT) as the basis to finance such an enormous undertaking. I won't pretend to have an answer as to how we will fund this global move to higher ground, but we have no choice. More than any other issue in human history, rapidly rising sea level will challenge us to think big at all levels—personally, professionally, locally, nationally, and globally.

What I do know is that the sea will rise to levels we now consider unimaginable. While questions of funding and priorities are relevant to us, they are irrelevant to the sea, which will work its way higher as the ice melts. The sea cares not about our budget challenges. The "glass half full" viewpoint from Chapter Nine points to the possibility that this problem can become an opportunity, an engine for growth. A quote from Goethe exhorts us to think big: *Whatever you can do or dream you can, begin it. Boldness has genius, power, and magic in it.*

11

THE PATH FORWARD

We are now faced with the fact that tomorrow is
today. We are confronted with the fierce urgency
of now. In this unfolding conundrum of life and
history, there is such a thing as being too late.
This is no time for apathy or complacency. This
is a time for vigorous and positive action.
—Martin Luther King Jr.

So, what now, you ask? Good question. While there are no easy solutions, there is a path forward, and a role for each of us to play.

Some readers will recall underwater explorer and pioneer Jacques-Yves Cousteau, the inventor of scuba diving. From the 1940s to the early 1990s, he was a global leader of environmental awareness, vastly expanding our understanding of our planet. Jacques had a towering influence, with worldwide recognition. *Time* magazine recognized him as one of the hundred most influential people of the twentieth century. His prominent series of television specials, *The Undersea World of Jacques Cousteau*, brought the oceans into people's living rooms across the globe. For a half century he was a singular communicator of science to global political leaders and directly to the public.

Months before Cousteau died in 1997, he and I spent a few days together when he hired me to become CEO of the Cousteau Society. Though my time with him was brief, he had a profound impact on my life and my thinking. He believed that citizens and organizations should take the initiative and advocate solutions. Frankly, he was quite suspicious of most modern political leaders. He said that if we, as concerned citizens and organizations, identify issues and build support, the politicians will want to "join the parade." That wisdom is worth considering as we ponder what each of us can and should do. His point was that we should not sit back and wait for politicians to lead. He advocated that we should create movements from the bottom up.

With Cousteau's guidance in mind, it is up to us to make sure that sea level rise gets the attention it deserves. The first step is to educate ourselves and others about the facts, and continue both efforts over time. How we distinguish fact from belief is a peculiar and important characteristic of our species. Each of us has a large amount of information that we consider facts, with varying degrees of clarity and certainty. We also have beliefs. Generally, we think we know the difference, but some beliefs become even more strongly held than facts.

It's hard to grasp the facts of unstoppable rising sea level for several reasons: It is without precedent in the human experience; it will be unimaginably disruptive; and, perhaps most importantly, there is a yawning chasm between the slowness of the perceived change and the urgent need for bold and durable solutions. We need to recognize that our beliefs—our desire to believe that this will not affect us—are standing in the way of the facts. Our biggest challenge has as much to do with our beliefs and emotions as with the facts at our disposal. Only by acknowledging that head-on can we begin to overcome our bias toward complacency. The case for rising seas is actually not even new. In addition to a century of warnings in various scientific publications, the highly popular Bell Telephone Science Hour television program very clearly made the case back in 1958 that continued burning of fossil fuels would warm the planet, melt the ice sheets, and raise sea level, putting coastlines underwater. (Spend two minutes looking at the actual video clip, which is fascinating.[40])

For over a half century, consciously or subconsciously, "we" put the forecast of rising sea level out of our minds. We did not want to abandon the benefits and convenience of fossil fuels. We liked the beaches, coastlines, and ports right where they have been throughout human civilization. Now with our heavily developed coastal cities, the idea of elevating or relocating vast structures and infrastructure is so difficult, disturbing, and disruptive that we simply turn our back on it and pretend it's not real. We are letting our beliefs and desires color our view of the facts. Despite knowing better, we continue to let ourselves believe that SLR is a problem for the distant future, to be left for the next generation, enabling us to avoid action today. As with any vexing problem, the temptation is to kick the can down the road and hope that some magical solution will appear.

TINY TANGIER

Consider this poignant example from historic Tangier Island, a little north of Norfolk, Virginia. Tangier is rapidly sinking into Chesapeake Bay, following the path of Sharps Island, which disappeared in 1962, and Holland Island in 2010—whose last house is shown in the next photo.

The rising water on Tangier is now impossible to miss due to a combination of land subsidence and rising sea level, adding to erosion. The tiny island of crabbers and fishermen has far less than a thousand people now due to the exodus. Each year, the flooding is getting visibly worse. In June 2017 a CNN television crew visited, meeting with the mayor and interviewing residents about the harsh reality that their community is rapidly disappearing just as the two nearby island communities did.

Figure 12. The last house on Holland Island in the Chesapeake Bay. The owners held on until 2009, a year before it fell into the bay. (Photo Credit: baldeaglebluff.)

President Trump saw the television program and called the mayor, James Eskridge, who came in from crabbing to take the call. As reported by both the mayor and the White House, they discussed the flooding and agreed it was just normal erosion, not sea level rise. The president and the mayor's shared belief ignored the factual findings of the US Army Corps of Engineers and NOAA, the National Oceanic and Atmospheric Administration, which both cite the phenomenon of rising sea level as a significant factor that is getting worse. According to the mayor, the president ended the call telling him:

> . . . not to worry about sea level rise. Your island has been there for hundreds of years, and I believe your island will be there for hundreds more.

Setting aside any politics, I find this story of tiny Tangier Island illuminating. Even in a small tightly knit community of *watermen*, as they are known locally, facts can be set aside in favor of beliefs. Erosion and subsidence are natural flooding progressions, similar to Seattle's in our opening example. What is different in recent decades for Tangier Island and nearly all coastal communities is the visibly increasing rate of global sea level rise. Even though it's by just fractions of an inch each year at present, the amounts tie in with the increasing melt rates of the glaciers and ice sheets on Greenland. The science is remarkably clear. We have passed a tipping point. Sea level is rising and accelerating very quickly. Hopefully we can slow the warming before it goes exponential. Regardless, we must begin real adaptation in the face of the rising sea, without delay.

A decade ago, when I was writing *High Tide on Main Street*, the book's title was a metaphorical futuristic threat. It is now a routine occurrence in the town of Tangier and a growing number of coastal communities all over the world. Sooner or later all of those residents will be moving to higher ground. We now know enough to see the path forward.

Two thousand years ago, Julius Caesar wrote, *"Men believe what they want to believe."* It seems just as true today. Most of us cling to our beliefs, either those that benefit us or those that fit with our ideology or philosophy. Resistance to reality can be surprisingly strong when reality goes against an emotional belief. With few exceptions, our species is guided by our beliefs and biases as much as we are by facts. In the foreword, Senator Angus King cited the *recency effect*, the tendency to assume that the future will be like the recent past. *Optimism bias* describes our assumption that things will turn out well "somehow." *Confirmation bias* is the term for giving attention to the evidence that supports the case we prefer

The problem is the same human tendency that prevailed when Galileo, Copernicus, and others made the case that the Earth revolves around the Sun. It took decades for that idea to be accepted. Even many scientists clung to their *beliefs* that the Earth was the center of the universe. There was fierce, vitriolic debate. Galileo was famously imprisoned by the Catholic Church for heresy. Although most of us do not have to fear imprisonment for our views, we are in a comparable era with regard to accepting the realities of climate change.

In sharp contrast to their position with Galileo, the Vatican's 2011 report on climate change and sea level rise, "Fate of Mountain Glaciers in the Anthropocene," was excellent and on target about this scientific truth. Just seventeen pages and easy to read, its explanation of the problem of rising sea level and recommended actions are excellent. Download it at www.movingtohigherground.com/vaticanreport.

To give another, more recent, example: As previously cited, glaciologist Dr. John Mercer forecast in 1978 that within fifty years West Antarctica's glaciers could start to thaw, advance into the sea, and eventually raise global sea level by 5 meters (16 feet). Hardly anyone believed it was possible; hardly any scientists took him seriously. Now it looks like he was remarkably close about the accelerating destabilization and collapse of West Antarctica, and with the ominous message about rising sea level.

Again and again, despite the best efforts of leading scientists, we seem inclined to turn our backs on the facts believing that it will all magically get better. In the present day, fortunately, we have our own generation of iconoclast scientists able to step outside the "norms." Physicist Dr. James Hansen and glaciologists Drs. Eric Rignot and Jason Box exemplify the ability and willingness to say that scientific caution and reticence may be preventing us from seeing the big-picture threat from SLR; certainly there are others with a similar posture.

One special source of inspiration is Greta Thunberg, the Swedish teen who has become a leading voice for bold action. She has made a name for herself scolding adults, telling everyone to listen to the scientists, and demanding her generation's right to a livable planet. Her succinctness and tone have elevated her as a world leader. She joins numerous advocacy organizations[41] who see the rise of carbon dioxide, global temperature, and sea level as existential threats to our way of life. I am in awe of her clarity and communication. She is a model for others. Greta knows that climate change is going to be the defining issue of her generation. It is not *a* political issue. It is *the* issue. Her generation's future depends on our willingness to make this a priority above all else, starting *now*. Greta and her generation are putting facts before denial, reality before wishful thinking. I believe that older generations can learn from her and those like her, who are unafraid to sound the alarm on the precarious state of the planet and make everyone aware of their role in fixing it. Greta sets a great example of belief in science and getting her facts straight.

We must frame rising sea level as our legacy, as it will reshape the world effectively *forever*. Stabilizing the world we leave our children and grandchildren will be a strong motivator for many. They will likely turn to us in twenty or thirty years and wonder what we did to plan for the rising sea and shorelines moving inland. We should have an answer. If we start today, there is still time to find innovative solutions that will protect assets, help humanity, and enable us to serve as pioneers in this challenging new era. I strongly believe that rewards will go to those who innovate new ways to operate in the coastal zone. We can do this with foresight for all those who follow us, our "children" in the broadest use of the term. We can do it because there is an economic opportunity for the far-sighted entrepreneur. We can do it as leaders with a strong moral compass. The key is to find a motivation that enables us to overcome our blind spots and face the reality head-on: to recognize the scope of the challenge, to set our sights on the future, and to begin.

TITANIC: METAPHOR AND LESSON

In *High Tide on Main Street*, I used the RMS *Titanic* as a metaphorical way to consider our options with regard to sea level rise. As is well known, the famous cruise ship emerged from a fogbank to discover a large iceberg in its path. Collision was unavoidable, which was my metaphor for the catastrophe posed by rising sea level. I said there were five options for how each individual might react: "partying," being paralyzed with fear, finding the responsible party, preparing for impact with efforts to minimize the effects, or considering what to do following the impact. The five choices apply to us with the climate crisis.

I recently discovered a stunning *Titanic* analysis in London at the Institute of Marine Engineering, Science and Technology (IMAREST) that's worth sharing. It showed that the captain had another option that would have likely allowed all the passengers and crew to get off the ship onto the iceberg, able to await rescue. Rather than turning hard to port (left), putting the engines into reverse to try to avoid hitting the iceberg, what if the captain had turned the ship somewhat to starboard, to intentionally ram into the iceberg, head-on, under power, jamming the vessel into the ice? The detailed evaluation showed that it would have kept the ship afloat until rescue ships arrived. The study was done by the Royal Institute of Naval Architects working with Harland and Wolff—the Irish company that built the ship.[42] My takeaway is that good solutions are not necessarily obvious, nor easy. They require good information, smart thinking, and leadership. Likely they will not get consensus support; they may even struggle to get majority support from the public, and even from the scientific and engineering communities. I like this example because it suggests that we increase our chances of survival by facing the obstacle in our path, head-on.

Sea level rise is a challenge without precedent. As I've described in this book, it will require exceptional leadership at all levels: innovative thinking, creative solutions, and tremendous vision. But to start, I believe it requires that we overcome our denial and wishful thinking and get the facts out into public discussion. Yes, we need great leaders at the local and national levels. Yes, we need business experts and engineers and architects. But that does not mean you can't be a change agent. In fact, what we need *most of all* is for all of us to start an international conversation about the rising ocean. To use the popular idiom, we need to "socialize" the concept that sea level rise is just getting started, and that we need to plan a path forward while there is time for us to get ahead of it. We need to go to community meetings. We need to start getting vocal about this issue so that leadership takes notice and gets to work getting the right teams in place to study appropriate solutions for each community.

Facing the facts is the first step. That helps harness our passion for solving this crisis, which will create the momentum we need to overcome it. That means identifying the things that people care about that will be affected by SLR and using that to motivate them to action. Humanitarian concerns will loom large. For some, leaving a livable planet to their grandchildren will be the focus. For others, that will mean protecting their greatest assets for their children. Still others will see this as an unparalleled economic opportunity that will drive innovation. For others it will be professional innovation in architecture, civil engineering, finance, law, and other professions.

One of the strongest ways to overcome our inertia is to band together with others and begin to build momentum. If you talk to a neighbor, and they talk to a neighbor, and then you have a community meeting about it, and then local business leaders and officials get involved, we can build a groundswell of change; perhaps a tsunami is an appropriate metaphor. So that is where we must begin: Ask others what they know about this. Correct misinformation. Ask questions about what your community is doing. Hold a meeting. Send out links to credible articles. Asking local leaders about their plans may be more effective than telling them what to do.

If you take one thing away from this book, it should be that you have the power to build understanding and urgency around this issue. Regardless of what the timing turns out to be, sea level rising more than 10 feet (3 meters) higher will be the biggest physical change to our planet in all of human civilization. Our challenge is to turn toward it, not away from it. The steps we take now can be our legacy. Whether you share this book, use the slides I am offering freely, prefer other references, or develop your own material, please make this your issue too.

Cousteau's advice that I cited at the start of this chapter continued with an insight that is relevant. To tackle something and get it on the public agenda, he told me, it should have three attributes. First, even complex issues had to be presented in a relatively simple form. Second, it had to be visual. Third, it had to be emotionally compelling. (He successfully used the approach to stop France's nuclear testing in the Pacific and to get approval for a treaty protecting Antarctica.)

Rising sea level has those three features. The melting polar ice caps, causing the sea to rise and move global coastlines inland, is clear, visual, and emotive. Though important on its own, it can be a lever to deal with the larger, more complex issues of climate change, our relationship to planet Earth, and how we will govern ourselves as we move into this profoundly different era.

So let us begin, to "rise with the tides" . . . for ourselves, for our moment on Earth, but more importantly, as testament of our concern for future generations.

We are not here to curse the darkness, but to light a candle that can guide us through the darkness to a safe and sure future. For the world is changing. The old era is ending. The old ways will not do. The problems are not all solved and the battles are not all won and we stand today on the edge of a New Frontier—a frontier of unknown opportunities and perils, a frontier of unfulfilled hopes and threats.

—John F. Kennedy

AFTERWORD

SIR DAVID KING

The climate emergency is like nothing that human civilization has ever faced. I have now worked on aspects of climate change for decades, dealing with the science and policy as chief scientific adviser to the UK government, and on international diplomacy through my position as lead on climate change negotiations for the Foreign Secretary in the UK.

It is of grave concern that our societies misunderstand or at least underestimate the seriousness of the climate problem, which is surely now a crisis. We are at the point of having multiple challenges simultaneously. We have to produce sufficient energy while drawing down the level of greenhouse gases. We need more land for agriculture to meet the demands of the growing world population. We need to deal with the rising risk of deluge rain, severe droughts, and extreme fires. Shrinking glaciers will affect drinking water supplies and agricultural productivity, particularly in Asia. This awareness should push us to work harder to slow the warming and damage as soon as possible.

However, because we avoided the difficult decisions for so long, now we must try to preempt disaster with aggressive measures to repair the global climate system and refreeze the poles and the Himalayas. These are the ambitious objectives of the Centre for Climate Repair, which I have established at Cambridge University. Hopefully we will succeed.

Yet, even if we get it all right, we are going to have to adjust to higher sea level in the coming decades and centuries. Greenland and Antarctica will melt for a long time despite our best efforts. John Englander clearly makes that case in this excellent book. As he explains, due to the excess heat already stored in the sea, we are committed to many meters of higher sea level, which will have a profound and permanent impact on coastal communities everywhere. We have passed a "tipping point." There should be no delay in beginning to adapt in advance of the rising sea, which will have the effect of shifting most shorelines inland.

Moving to Higher Ground helps us to see our ocean planet differently. Indeed, the planet will soon look somewhat different when sea level is meters higher. This book gives us fair warning of the profound change coming to all coastal areas. It gives us time to prepare, but we must get on with it.

University of Cambridge
June 2020

ACKNOWLEDGMENTS

Books may not be sentient beings, but they are complex, organic creatures. Finding the roots and giving due credit to those who contribute to their genesis, development, character, and quality is not easy.

Moving to Higher Ground certainly is built on the foundation of my previous book, *High Tide on Main Street,* so all of its acknowledgments apply. As I explained in that preface, my decision to write a book about unstoppable rising sea level had a clear moment of genesis. It was an epiphany during my first night in Greenland, in August 2007. I remember the spot and the moment it happened and have been back there many times.

I had no idea what to expect with that first book, but was gratified with the near-universal acceptance and reception it received when first published in October 2012. In the years since, I have given hundreds of presentations to diverse audiences, including some that are highly expert. Their questions and comments led to this new work. Kevin Donovan, Will Rowe, Jr., Don Bain, Alec Bogdanoff, Lisa Craig, Jim Murley, Will Travis, Daniel Kreeger, David Punchard, and Mitch Chester provided strong encouragement for this project.

Moving to Higher Ground attempts to deal with the obstacles to action as we absorb the stunning reality that global sea level will be at least several feet (or meters) higher than it has been during all of human civilization. This book required thinking more about the psychology and social aspects as we absorb this change into our lives, communities, and professions. These aspects came from dozens of interactions, impossible to cite. Nonetheless, thank you to all who confessed your disbelief or anxiety about the message I was presenting. You provided insight, inspiration, or ingredients that are in this book.

The actual direction and guidance for the book came from a much smaller group of people, both those who offered feedback over the last eight years as the book took shape and those who reviewed drafts of the manuscript. At the risk of missing some, the following gave significant input on the content, both factual and stylistic: Bob Corell, Gary Griggs, Mike Shatzkin, Peter McCarthy, Katherine Flannery, Phil Turner, Neil Ross, John Booth, Jill Buchanan, Tim Fox, Jeff Onsted, Judy Keiser, and my wife, Linda Englander. Sharon Gray, my research assistant, was very helpful on the Deeper Dive Notes, the supplemental online material. Fortuitously, Bill Gleason, a long-time friend and highly experienced publisher, came out of retirement to take over The Science Bookshelf, and expand the distribution of this book.

Dozens of scientists and engineers answered questions, which helped me work out how to explain the scenarios for sea level rise. Climate experts Drs. James Hansen and Makiko Sato provided the components of what is now my "signature graph" in the first chapter.

Having Senator Angus King and Sir David King (no relation) take the time to write the foreword and afterword respectively is a great honor and should add interest and credibility to this work. Jim Bertram's cartoons at the start of each section are brilliant.

For this book, my publisher, the Science Bookshelf, contracted with Girl Friday Productions for all the editing and production, which they did with great professionalism. Their entire team, led by Sara Addicott, was invaluable and made this book what it is. In particular, Christina Henry de Tessan and John Peragine each did substantial editing, helping me find the right flow and balance of science so that the resulting work might have a broad audience. Justine Bylo of Ingram Publishing Services and Marissa Eigenbrood of Smith Publicity were strong advocates for this book.

Several friends supported my work and the publication of this book in particular: Carl Eibl, Michael Goldstein, Gary Greenberg, J. Mark Grosvenor, Jan Kruthoffer, Frank Rodriguez, and Tim Schweikert; also, the Serendipity Foundation.

Finally, my wonderful wife, Linda, and daughter, Rachel, have endured the added years of my being absorbed by this work. They have been very indulgent of what can only be described as my obsession to fully understand and tell this story.

Thank you all.

OTHER RESOURCES

REGARDING SEA LEVEL RISE

By John Englander

- Slides to teach rising sea level to general audiences: www.johnenglander.net/publicslides

- Full presentation: On YouTube "John Englander Royal Institution"

- "Sea Level Rise Now," a weekly blog and news digest. www.johnenglander.net

Websites

- US Government www.climate.gov Good visualizations of sea level and other climate related data

- www.nasa.gov search "sea level"

- https://earthobservatory.nasa.gov/topic/snow-and-ice

- National Snow and Ice Data Center for latest about Greenland and Antarctica www.nsidc.org

- US National Climate Assessment globalchange.gov/climate-change

- Georgetown Law Center's "Managed Retreat Toolkit" www.georgetownclimate.org

- Sea Level Rise www.sealevelrise.org Consumer level public infographics

- "Surging Seas" information from Climate Central sealevel.climatecentral.org

- NOAA's Digital Coast https://coast.noaa.gov/digitalcoast/

- Flight of the Frigate Bird, www.flightofthefrigatebird.com, a beautiful video blending science and story about vulnerable Dauphin Island, in a cautionary tale.

Books

- *A New Coast* by Jeffrey Peterson. Analysis and recommendations for US Coastal Policy

- *A Farewell to Ice* by Peter Wadhams. Powerful and poignant tale of Arctic melting

- *Adapting Cities to Sea Level Rise: Green and Gray Strategies* by Stefan Al. Concepts for architecture

- *Adapting to Rising Sea Levels: Legal Challenges and Opportunities* by Margaret Peloso

- *Rising: Dispatches from the American Shore* by Elizabeth Rush. A series of wonderful descriptive essays

- *Sea Level Rise* by Orrin Pilkey and Keith Pilkey. Similar focus, by coastal experts

- *The Long Thaw* by David Archer. The fundamental science of global warming and climate change

- *The Water Will Come* by Jeff Goodell. More stories and narrative; very well-written

KEEPING YOUR ASSETS
ABOVE WATER CHECKLIST

TO DOWNLOAD:
www.movingtohigherground.com/checklist

Rising sea level dramatically increases the risks of traditional flooding. This checklist will help you assess your vulnerability and options. The source is the information developed and explained in *Moving to Higher Ground: Rising Sea Level and the Path Forward* by John Englander (The Science Bookshelf 2021). This checklist will help you assess your property and flooding risk.

Primary asset of concern: _____

Completed by: _____

Date: _____

CRITICAL INFORMATION CHECKLIST

1 _____   Depending on your location, you may be able to get a free flood risk report for your property from www.floodfactor .com or www.surgingseas.com. In the United States, good sources of information about property elevation and historical flooding for your area may be available from the US Geological Survey (USGS), US Army Corps of Engineers, FEMA, or NOAA websites.

2 _____   Know the minimum height of your property *above sea level* (ASL). You can determine this from a property survey.

Elevation phone apps such as Altimeter and computer apps may give you an approximation.

3 _____ Pay particular attention to the distance from water level to your lowest opening, e.g. doors, windows, and vents in the foundation.

4 _____ When assessing your property's vulnerability, I suggest allowing for a margin of safety of at least 3 feet (1 meter) of higher sea level by midcentury in addition to the highest historical flooding at your location. Ideally the margin of safety would be 10 feet (3 meters) over historic high water. Maps of historic flooding, such as the 100-year and 500-year floodplains, are good information, but with the warming planet, so-called "100-year flood" events can now happen far more often, even annually.

5 _____ Consider the vulnerability of your community. For example, your home might be 10 feet ASL, but if access roads and critical infrastructure like water supply and wastewater treatment are vulnerable to rising sea level and increased flood events, you are too.

6 _____ In addition to flooding from below, if there is terrain above you, also evaluate flooding from above in the form of water coming downhill from a higher-elevation property, down a ravine, or from a stream, river, or dam that might overflow with the increasing rainfall patterns. Flood patterns are rapidly changing due to increased rainfall caused by warming oceans and greater evaporation. Pay more attention to flooding of the last decade and to any trends in flooding.

7 _____ Remember that flooding can come from four potentially sudden primary sources: (a) coastal storm surge, (b) direct rainfall, (c) downhill or downstream runoff, and (d) extreme high tides, which can all occur in combination. Rising global sea level is a slow, unstoppable phenomenon

that will raise the base water level and the height of the other short-duration flood sources.

8 _____   Distance inland is not protection if the ocean has a pathway to your property such as navigable waterways or porous rock, marshland, or swamp. In coastal areas, storm surge may become even higher when it is trapped in bays, harbors, and inland waterways. For example, Sacramento is eighty miles inland but is at a low elevation on a tidal river and is just as vulnerable as San Francisco.

9 _____   Be aware that if you are in an area of land subsidence (the land is sinking)—e.g. the New Orleans region; Norfolk/ Hampton Roads; Venice, Italy; or Jakarta—you are exposed to sea level rising faster than the global average. In high latitudes like Alaska, Canada, Russia, and Scandinavia, local sea level may be falling due to land uplift (land rising), but within a few decades that will likely reverse, greatly surprising most.

10 ____   Rivers like the Mississippi in the US, the Thames in the UK, and the Elbe in Germany have complex flood characteristics. At the mouth they are tidal, with water level affected by the daily tides. In those areas, anticipating sea level rise is very relevant. Further upstream, at higher-elevation parts of the river, they are well above any imminent sea level rise.

11 _____   Lake properties are generally disconnected from sea level rise, though you still need to look at flood risk from heavier-than-normal rain and runoff, related to warming oceans. Property on lakes and rivers may actually increase in relative value as sea level continues to rise and people move away from the coasts but still want to live on the water.

12 _____   Consider the time horizon for which you want to keep the property.

- If less than ten years and the property isn't currently experiencing flooding from extreme high tides, then the concern from potential sea level rise is much less. You should, however, keep in mind the coming damaging effects of increasing flood insurance costs and the market's growing awareness of increased flood risks. A big factor during the next ten years will be banks' willingness to offer thirty-year mortgages for properties vulnerable to even minimal sea level rise. These factors could all hamper your ability to sell within the next ten years.
- If your time horizon is between ten and twenty years, you should evaluate the risks of sea level being 6 to 12 inches higher. This may not sound like much, but keep in mind that extreme high tides will stack on top of this new base height; many wastewater and freshwater treatment systems will be negatively affected. Within this time frame, all the risk factors listed above, as well as insurance costs, mortgages, and public perception, will impact real estate values.
- If your time horizon is greater than twenty years, you must consider the impacts on your property and surrounding community of sea level being potentially 1 to 3 feet higher. This will be devastating for many low-lying coastal communities. This time horizon may seem far in the future, but keep in mind that twenty-five years ago it was the 1990s, and that doesn't seem that long ago.

13 _____   Consider your age and how you want to use the property.

- If you are young and plan to raise your children in the home, it will be important to have a long time horizon.

•   If you are older, consider the need to floodproof your home if you're planning to stay in it for more than ten years, especially if you are considering leaving it for your children or grandchildren. If you plan to relocate, it is usually preferable to make such a move before old age makes it difficult to do.

14 _____   Consider your risk tolerance to property devaluation. If the value of the property is a large part of your net equity (investments), you are more vulnerable and you may want to reduce your risk exposure. If, on the other hand, you have considerable other equity and investments, even an expensive home in a flood-vulnerable area might be acceptable if you can afford the risk. Property values will likely go underwater long before the property actually does.

15 _____   Evaluate your flood insurance. Even if you presently have flood insurance, the cost is likely to increase in the years ahead, potentially dramatically. A significant increase in the cost of flood insurance will have a significant impact on the value of a property. By one estimate, a five-hundred-dollar increase in the cost of flood insurance could reduce the value of a typical home by ten thousand dollars. In some vulnerable locations, flood insurance may not be available in the future or may be unaffordable.

16 _____   If your property has substantial value, you may want to consider having a professional vulnerability assessment done. Local architects or engineers should be able to refer you to someone who offers such a service.

NOTES

1   The four-minute trailer, *Chasing Ice Official Video*, is available on YouTube.com.

2   Mercer, J. H. (1978). West Antarctic ice sheet and CO_2 greenhouse effect: a threat of disaster. *Nature*, 271(5643), 321–325.

3   "Rising Seas: Record Warmth Found at 'Doomsday Glacier' Water Line." *ExtremeTech*. https://www.extremetech.com/extreme/305718-rising-seas-water-thwaites-glacier-grounding-line-above-freezing.

4   See my blog post, "Radical Scheme to Stop Sea Level Rise?" at www.johnenglander.net/blogs for my explanation why this is completely unrealistic. There you'll also find a link to an article in the *Atlantic*.

5   Allison, I. et al. (2009). Ice sheet mass balance and sea level. *Antarctic Science*, 21(05), 413-426. Retrieved from http://journals.cambridge.org/abstract_S0954102009990137

6   K. Lambeck, et al. 2014, "Sea level and global ice volumes from the Last Glacial Maximum to the Holocene", Proceedings of the National Academy of Sciences of the United States of America, 111(43), 15296-15303, doi: http://dx.doi.org/10.1073/pnas.1411762111

7   John Englander, *High Tide on Main Street* (Boca Raton: The Science Bookshelf, 2012), 31-32.

8   See www.sciencealert.com, search for: 5 atom bombs

9   Sweet, W.V., R.E. Kopp, C.P. Weaver, J. Obeysekera, R.M. Horton, E.R. Thieler, and C. Zervas, 2017: *Global and Regional Sea Level Rise Scenarios for the United States*. NOAA Technical Report NOS CO-OPS 083. NOAA/NOS Center for Operational Oceanographic Products and Services

10 Ibid., 22.

11 D. Archer and V. Brovkin. 2008. "The Millennial Atmospheric Lifetime of Anthropogenic CO_2." *Climatic Change* 90(3), 283–97.

12 O.A. Dumitru, et al. 2019. "Constraints on global mean sea level
    during Pliocene warmth." Nature, 574(7777), 233–236. https://
    doi.org/10.1038/s41586-019-1543-2

13 J. Bastin, Y. Finegold, , et al. 2019. The Global Tree Restoration
    Potential. Science 365(6448), 76–79. DOI: 10.1126/science.
    aax0848.

14 https://missionaransas.org/resilient-texas-planning-sea-level-rise

15 For more about the fascinating and relevant extreme tides, I rec-
    ommend the book *Tides: The Science and Spirit of the Ocean* by
    Jonathan White (San Antonio: Trinity University Press, 2017).

16 https://www.nytimes.com/2019/06/19/climate/seawalls-cities-
    cost-climate-change.html?smid=nytcore-ios-share

17 https://www.weforum.org/agenda/2019/01/the-world-s-coastal-
    cities-are-going-under-here-is-how-some-are-fighting-back/

18 See https://www.imeche.org/policy-and-press/reports/detail/ris-
    ing-seas-the-engineering-challenge, or use this easier URL: www.
    movingtohigherground.com/imeche.

19 www.bostonlivingwithwater.org/

20 "Notice of Intent to Prepare a Draft Environmental Impact
    Statement for the Lake Pontchartrain and Vicinity General
    Re-Evaluation Report, Louisiana," *Federal Register* 84, no. 63,
    Tuesday, April 2, 2019: 12598, Notices.

21 "New Research Analyzes Countries at Greatest Risk from Climate
    Change Impacts" (2007), https://www.earth.columbia.edu/
    articles/view/958.

22 https://www.theguardian.com/cities/2015/feb/19/
    thames-barrier-how-safe-london-major-flood-at-risk

23 A. Ali. 1996. "Vulnerability of Bangladesh to Climate Change and
    Sea Level Rise through Tropical Cyclones and Storm Surges."
    Water, Air, & Soil Pollution 92(1–2): 171–79. doi:10.1007/
    BF00175563.

24 For more information, https://www.shidhulai.org/ and https://
    www.alizecarrere.com/adaptation

25 P. L. Bernstein. 1998. *Against the Gods: The Remarkable Story of
    Risk.* Hoboken, NJ: Wiley. 6–7.

26 https://www.genevaassociation.org/sites/default/
    files/research-topics-document-type/pdf_public//

ga2013-warming_of_the_oceans.pdf

27 A. G. Baribeau. 2018. "The SLR Factor—As Sea Levels Rise, the Flood Risk Equation Changes." Actuarial Review Magazine 45(1), 18–24. https://ar.casact.org/magazine_issues/march-april-2018/.

28 A good example is "The Problem with Levees" by Nicholas Pinter in *Scientific American*, August 1, 2019, https://blogs.scientificamerican.com/observations/the-problem-with-levees/.

29 See my book *High Tide on Main Street*, pp. 110, 112.

30 http://www.freddiemac.com/research/insight/20160426_lifes_a_beach.page

31 A. Bernstein, M. Gustafson, and R. Lewis. 2018. "Disaster on the Horizon: The Price Effect of Sea Level Rise." *Journal of Financial Economics*. May 2018. https://ssrn.com/abstract=3073842 or http://dx.doi.org/10.2139/ssrn.3073842.

32 https://www.wsj.com/articles/flooding-risk-knocks-7-billion-off-home-values-study-finds-1535194800

33 "The Risky Business Report" by former US Secretary of the Treasury Hank Paulson, former New York City mayor Michael Bloomberg, and hedge fund investor Tom Steyer.

34 "One way or another the deluge is coming" *The Economist*, August 17, 2019. https://www.economist.com/leaders/2019/08/17/one-way-or-another-the-deluge-is-coming

35 March 25, 2019, https://www.frbsf.org/economic-research/files/el2019-09.pdf

36 "How to Reject Climate Porn and Reach Climate Acceptance," October 2, 2018, https://www.greenbiz.com/article/how-reject-climate-porn-and-reach-climate-acceptance

37 https://www.dezeen.com/2020/02/14/nothern-european-enclosure-dam-climate-change/

38 Download the PDF from www.johnenglander.net/SALT or from ISGP's website, www.scienceforglobalpolicy.org.

39 Accounts of the meetings are directly from the notes of my colleague Dr. Robert W. Corell.

40 www.johnenglander.net/1958TV

41 Examples of advocacy organizations are: Citizens Climate Lobby, 350.org, and Extinction Rebellion.

42 C. Hackett and J. G. Bedford. 1997. "The Sinking of the

SS *Titanic*—Investigated by Modern Techniques." *RINA Transactions and Annual Report.* Doi: 10.3940/rina.sbt.1997.b10.

INDEX

The abbreviation "fig." following a page number denotes a figure.

ABOUT THE AUTHOR

John Englander is a renowned oceanographer, multi-book author, speaker, and expert on climate change and sea level rise. His 2012 book, *High Tide on Main Street: Rising Sea Level and the Coming Coastal Crisis*, explained the science in easy-to-understand terms. *Politico* listed it as one of the top fifty books to read. It continues to be the best-selling book on the topic.

His broad marine science background, coupled with explorations in Greenland and Antarctica, allows him to see the big-picture impacts of changing climate and rising seas on society. Millions of people in the US and around the world have read his books and blog posts, or heard his popular talks. Without any political bias, he explains the science in plain language, often using personal anecdotes. Conference organizers, military leaders, and various professional societies are on the record rating Englander as the best speaker on the subject.

Englander works with businesses, communities, and government agencies to understand why sea level will rise far higher than most can imagine. Increased flooding comes with the compounding effects of rising seas, extreme tides, and severe storms. He advocates for "intelligent adaptation." Along with the tremendous risks in the coming decades, he believes there will also be enormous economic opportunities that will allow us to thrive if we begin to plan and adapt *now*. He is the founder of a new nonprofit, the Rising Seas Institute.

Englander brings the diverse points of view of an industry scientist, entrepreneur, and CEO. For more than four decades, he has been a leader in both the private and nonprofit sectors, serving as chief executive officer for such noteworthy organizations as the International SeaKeepers Society and the Cousteau Society. For more than two decades he was a professional in the diving industry, logging several thousand dives, including leadership of two expeditions under the polar ice cap.

Englander is a sought-after subject matter expert with appearances on MSNBC, Fox Business, ABC, CBS, PBS, the Weather Channel, CCTV (China), CBC (Canada), NPR, and SkyNews TV (UK). He has been featured in *USA Today*, the *Huffington Post*, the *San Francisco Chronicle*, the *Miami Herald, Publisher's Weekly*, and numerous other publications.

Englander graduated from Dickinson College with a degree in geology and economics. He is a research fellow at the Institute of Marine Sciences at UC Santa Cruz; a fellow of the Institute of Marine Engineering, Science and Technology (IMarEST); a fellow of the Explorers Club; and a member of several professional societies. For more information about the author and access to his blog posts, visit www.johnenglander.net.

STAY IN TOUCH

WEEKLY BLOG/NEWSLETTER: For nearly a decade, John has published *Sea Level Rise Now*, delivering up-to-the-minute news on SLR and climate change.

Each week, you receive a link to John's most recent blog post and news of high interest to those looking for a clear analysis of what is going on in the world of climate change and SLR. Entertaining and informative, *Sea Level Rise Now* explores the cutting edge topics that will profoundly affect our futures.

To subscribe, visit https://johnenglander.net/blog.

The blog/newsletter is free. Your email address is strictly confidential and will not be shared or distributed anywhere else.

SOCIAL MEDIA

FACEBOOK: https://www.facebook.com/johnenglanderSeaLevelRise/
TWITTER: https://twitter.com/johnenglander
LINKEDIN: https://www.linkedin.com/in/johnenglander/
INSTAGRAM: https://www.instagram.com/johnenglander/

CONTACT THE AUTHOR

John Englander welcomes direct comments, insights, and emails from his readers: john@johnenglander.net

SPEAKING ENGAGEMENTS

John is a frequent keynote/expert speaker at conferences, sharing his expertise on climate change and sea level rise.

He is also available for small, medium, and large groups via webinar or teleconference. For more information, please visit:

https://johnenglander.net/speaking